W0234629

The AADHAAR Effect

The AADHAAR Effect

why the world's largest identity project matters

N.S. RAMNATH | CHARLES ASSISI

FOUNDINGFUEL.com

OXFORD
UNIVERSITY PRESS

OXFORD
UNIVERSITY PRESS

Oxford University Press is a department of the University of Oxford. It furthers the University's objective of excellence in research, scholarship, and education by publishing worldwide. Oxford is a registered trademark of Oxford University Press in the UK and in certain other countries.

Published in India by
Oxford University Press
2/11 Ground Floor, Ansari Road, Daryaganj, New Delhi 110 002, India

First Edition published in 2018

ISBN-13 (print edition): 978-0-19-948761-5
ISBN-10 (print edition): 0-19-948761-8

ISBN-13 (eBook): 978-0-19-909603-9
ISBN-10 (eBook): 0-19-909603-1

Typeset in Adobe Garamond Pro 11.5/16
by Tranistics Data Technologies, Kolkata 700 091
Printed in India by Nutech Print Services India

For my parents.

—N.S. Ramnath

For Dad, who taught Kolya and me to read, write, and live.

—Charles Assisi

Contents

Abbreviations

AePS	Aadhaar-enabled Payment System
AGI	Attorney General of India
APBL	Airtel Payments Bank Limited
API	Application Programming Interface
BHIM	Bharat Interface for Money
BJP	Bharatiya Janata Party
BPL	Below Poverty Line
CAG	Comptroller and Auditor General
CCA	Controller of Certifying Authorities
CCT	Conditional Cash Transfer
DBT	Direct Benefits Transfer
DIKSHA	Digital Infrastructure for Knowledge Sharing
ePOS	Electronic Point of Sale
GPS	Global Positioning System
GST	Goods and Services Tax
GSTN	Goods and Services Tax Network
IAS	Indian Administrative Service
IFS	Indian Foreign Service
IRS	Indian Revenue Services

iSPIRT	Indian Software Product Industry Round Table
KYC	Know Your Customer
LPG	Liquefied Petroleum Gas
MDR-TB	Multidrug-Resistant Tuberculosis
MFI	Micro-finance Institution
MGNREGA	Mahatma Gandhi National Rural Employment Guarantee Act
MSMEs	Micro, Small, and Medium Enterprises
NDA	National Democratic Alliance
NPCI	National Payments Corporation of India
NPR	National Population Register
NTP	National Teacher Platform
PDS	Public Distribution System
PMJDY	Pradhan Mantri Jan-Dhan Yojana
PMO	Prime Minister's Office
PNB	Punjab National Bank
RBI	Reserve Bank of India
RTI	Right to Information
SBI	State Bank of India
SMEs	Small and Medium Enterprises
TB	Tuberculosis
TRAI	Telecom Regulatory Authority of India
UID	Unique Identification
UIDAI	Unique Identification Authority of India
UN	United Nations
UPA	United Progressive Alliance
UPI	Unified Payments Interface

Acknowledgements

POPULAR NARRATIVE HAS IT THAT WRITERS ARE solitary creatures who live in worlds only they can imagine. After two years of relentless research and writing for this book, we feel compelled to assert that writers seek solitude only when they need time out to faithfully describe all that they have witnessed or heard. To interpret the truth embedded in the many stories, writers work closely with other people.

The fact is, this book may not have come to life if Indrajit Gupta had not spotted the significance of Project Aadhaar, impressed upon us the need to research and report it first-hand, scrutinized our assertions, and helped open doors that may otherwise not have opened.

Like Gupta, C.S. Swaminathan was among the early ones who saw the potential in this narrative. His no-holds-barred feedback basis the practical wisdom gained from years in business enormously helped us think through our arguments. It's comforting to know that he, too, was watching out for us while our heads were buried in research and writing.

Sveta Basraon meticulously edited multiple drafts of the manuscript. Sveta would morph from colleague into concerned friend when she thought it necessary that we call it a night so we may wake up to write another day.

When Kavi Arasu joined Founding Fuel, he started work right away as a coach to the founders. And while we were reporting and writing, he'd often step in with advice and exercises on what it'd take to complete the task even as we stayed cognizant of the nuances.

These nuances are of the kind that often raised philosophical dilemmas in our minds. At times like these, Arun Maira, with his sagacity and remarkable ability to listen deeply, offered pointers on how to grapple with them. He also told us some conflicts are impossible to resolve; that we must accept ambiguity and multiple world views with serenity, for which we are grateful.

While we engaged in hundreds of hours of one-on-one conversations with key stakeholders, Govindraj Ethiraj had interviewed them in the early days of the project when he was on a sabbatical from journalism and with the UIDAI. The generous soul that he is, he shared these rare audio recordings to help us. Similarly, our former colleague Mitu Jayashankar, who has tracked the project since it was thought up—and Nandan Nilekani's career for longer—shared her insights. M. Rajshekhar, who has done some of India's best reporting from the ground and tracked the project through its early days, also offered us his learnings. Without Govind, Mitu, and Rajshekhar's contributions, this book would have been the poorer.

Nandan Nilekani is central to this book and to Aadhaar. To understand his motives, why this project matters to India and the developing world, the nature of the current debate, and how he sees the future, it was crucial he let us into his mind. He engaged with us in deep conversations, shared key learnings and the mental models he deploys. After listening to him, we started to appreciate second-order thinking. While Nilekani let us into his mind, his wife Rohini Nilekani offered us pointers to understand his softer side. We are thankful to them for staying engaged with us over email, on the phone, and conversing with us at their workplaces and their home.

There is much we owe Shankar Maruwada, who co-founded the not-for-profit EkStep with Nandan and Rohini Nilekani. He was part of the founding team at UIDAI, is a close confidante of Nilekani, and has a ringside view of the man's mind. Most importantly, he possesses the remarkable ability to connect the dots. Our conversations with him helped us appreciate why identity matters to an individual. They cut through the noise that surrounds polarizing debates and made us think about what the future of platforms may look like. He pushed us as well to ask questions that lie at the intersection of technology, society, and philosophy.

Even as this note is being written in August 2018, Ram Sevak Sharma has received an extension as head of the telecom regulatory authority of India (TRAI), until 2020. We had heard him speak on the project with tremendous energy and passion back in 2010. His energy and passion was intact when we met him in 2017 at his office in New Delhi. We cherish these conversations for their transparency and openness.

Much like Sharma, Ashok Pal Singh, who now heads India Post Bank, K. Ganga, and Anil Kumar Khachi are career bureaucrats. Between them, they upturned all notions we had about bureaucrats. They were part of the founding team at UIDAI, and helped us understand the complexities of governance.

In A. Babu, another career bureaucrat, based in Andhra Pradesh, we saw what it takes to convert an idea into action, and are grateful to him and his team for showing us the challenges on the ground. We are also thankful to many other state-level public servants for showing us their rather thankless world.

Then there were those from the private sector who were part of the founding team of UIDAI. These include the project's chief architect Pramod Varma, product manager Sanjay Jain, and the head of technology Shrikant Nadhamuni. Each of them spent long hours taking us over how they thought the project through, and did not flinch from tough questions. Shrikant Karwa, who was into operations as a volunteer, provided global perspectives and helped open many doors.

The team that came together at UIDAI may not have gotten to work on Project Aadhaar without support from the political class. What was the government thinking when it gave them the go-ahead? P. Chidambaram and Jairam Ramesh, who were ministers in the UPA regime when this project was flagged off, took time out to speak with us individually and in public. This offered us glimpses into the thin lines a politician must walk and exposed us to the intellectual firepower the political class possesses.

When in office, politicians do not make decisions in isolation, but in the company of fine minds. This made itself obvious when we sought out Montek Singh Ahluwalia and Arvind Subramaniam. Ahluwalia advised former prime minister Manmohan Singh and met us at his home in New Delhi. Subramaniam, now serving his tenure as chief economic advisor to the current government, met us at his office in New Delhi. Both were candid on what is possible and what is not. We are grateful for the time they offered us.

In a democratic country it is inevitable that citizens ask uncomfortable questions. Many of these have come from Usha Ramanathan, Kiran Jonnalagadda, Anand Venkatanarayanan, and Srinivas Kodali. Their voices have been heard and the constitutional validity of the project was raised in the Supreme Court. What are the issues that concern them, and what is it they think citizens must watch out for? All of them articulated in no uncertain terms why they stand where they do. They continue to stay connected with us, for which we are glad.

Then there are those who believe it is in the larger interest of society that a project such as this must gain traction. Those who believe in it have banded together as iSPIRT. Even as this book goes to print, Sharad Sharma, Sanjay Anandaram, Venkatesh Hariharan, Nikhil Kumar, Tanuj Bhojwani, and Siddharth Shetty continue to stay connected and share insights on the world as they see it.

Project Aadhaar is complex to understand and insists we engage with not just people, but navigate multiple ecosystems at once. Haresh Chawla is among those few people who are adept at understanding the digital economy and India at once. We are glad we knocked often at his door.

Academic rigour is called for as well. We sought it between long walks through the streets of old Bengaluru in search of the perfect *dosai* and coffee while conversing with Professor M.S. Sriram of IIM Bangalore. The gentle Professor Rishikesha T. Krishnan, director of IIM Indore and member of the Justice B.N. Srikrishna Committee set up to recommend data protection laws for India, responded to our multiple emails and phone calls. The exuberant Professor Mohanbir Sawhney of the Kellogg School of Management was happy to entertain our calls despite the long distance and the time difference.

What implications might this project have on public policy? Bindu Ananth of the Dvara Trust informed us of the nuances we must scrutinize.

When a project of this kind starts to take shape, those who head businesses are among the ones with the mental muscle to attempt spotting opportunities. While researching this book, Arundhati Bhattacharya was nearing the end of her term as the chairman of State Bank of India (SBI). In spite of the many pressures on her, she spent time with us and shared why she thought Aadhaar was important for India's largest bank. Between her and the bank's chief technical officer (CTO) Dhananjay Mahapatra's insights, we witnessed what it takes to steer an oceanliner.

At the other end of the spectrum was Kishore Biyani, chairman of the Future Group, reticent and welcoming at once. He likes to play his cards close to the chest. Having said that, it was eye-opening to learn how somebody in business stays pragmatic in order to conserve cash while taking risks in the face of an uncertain future. Then there is Manish Sabharwal, the co-founder of Teamlease Services, who spotted the potential in Aadhaar when it was announced and helped us think through its role in creating and formalizing employment.

How does the technology embedded in this narrative look like to experts outside the project? Krishnakumar Natarajan, executive chairman at Mindtree Consulting, Murari Sridharan, CTO at BankBazaar, Ram Ramdas, co-founder of Herald Logic, and Ritesh Pai, chief digital officer at Yes Bank, offered perspectives on how to examine the architecture of Aadhaar.

When large entities such as these get involved in the project, there are hungry entrepreneurs waiting for capital to exploit opportunities. Sanjay Swamy at Prime Venture,

Ganesh Rengaswamy of Quona Capital, and Sahil Kini from Aspada Investments have been watching the ecosystem closely to evaluate it for investment opportunities. Conversing with them yielded insights into the kind of entrepreneurial ecosystem that may emerge out of India and how to examine its chances against the narratives of Silicon Valley in the US and China.

How does this project look to somebody from the outside and what lens do you look at it from? Between Santosh Desai's ability to unspool the mind of India, Rahul Matthan's appreciation of law, and Niranjan Rajadhyaksha's informed takes on the state of the economy, very interesting pointers to the future of Aadhaar and India Stack emerged. More clarity emerged from public policy analysts, especially Pranesh Prakash, who has been tracking the project right from the beginning.

Founding Fuel's advisory council, which includes Harsh Mariwala, chairman of Marico; Sanjeev Bikhchandani, founder of Naukri.com; Kiran Karnik, former president of Nasscom; Rama Bijapurkar, a widely respected thought leader on market strategy and the Indian economy; Uday Shankar, president of 21st Century Fox (Asia); D. Shivakumar, executive president of the Aditya Birla Group; and Analjit Singh, founder chairman of Max Group, have been constant sources of inspiration.

The Aadhaar Effect is the first book in a series around themes crucial for those in leadership roles. For it to come to life, though, Founding Fuel and its mission had to be stated. When we shared it with a few people, they placed their trust in the team and suggested the co-founders walk the talk.

K.V. Kamath was gracious enough to unhesitatingly share his time and thoughts, for which we are now the richer.

We also owe much to K. Ramkumar, Arup Mazumdar, Susmita and Subroto Bagchi, V. Vaidyanathan, Rajesh Srivastava, Vijay Bhat, Nitin Thakur, Vijay Jain, Vinod Janardhan, Rajiv Rajgopal, Srinivasan B., Shivashankar V.N., and Srikanth Venkatesan.

Once Founding Fuel decided that learning is the business it wants to be in, we needed live case studies. Sandeep Banerjee, Deepak Iyer, Upendra Namburi, Meeta Dullabh, Anantha Nayak, Bisawdeep Ghosh, Bhargav Dasgupta, Alok Agarwal, and Jerry Jose stepped in to provide case studies they thought pertinent to leaders. Project Aadhaar and the learnings that emerge from it added to the arsenal.

To understand leadership, though, leaders must open up. Among the first who did were stalwarts from the retail business, including B.S. Nagesh, Damodar Mall, R. Sriram, Arvind Mediratta, Govind Shrikhande, Amit Agarwal, Darshan Mehta, Amit Jatia, Anuj Puri, Rajendra Kalkar, Sadashiv Nayak, and Vivek Biyani.

Thank you Baba Prasad, Indranil Chakraborty, Sumit and Amrita Chowdhury, Hari Abburi, N. Dayasindhu, Sanjay Handu, Gourav Jaswal, Sachin Joneja, Soum Paul, Nitin Srivastava, Harsh Vardhan, and Pankaj Tibrewal, for coming on board to co-create the Founding Fuel platform and sharing your knowledge generously. Even as the content they were creating started to fall into place, many willingly chose to partner with a fledgling entity.

Priya Naik, R. Sukumar, Madan Padaki, Biju Dominic, Devangshu Dutta, Piyul Mukherjee, Sheetal Choksi, Ranjan Banerjee and Bhaskar Das, Vinay Hebbar, Tejaswini Adhikari, Neelima Mahajan and the Cheung Kong Graduate School of Business in Beijing, Dr Rajat Chauhan, Indranil Gupta,

Arun Kaushik, Ruubina Shah, N. Dayasindhu, Dipayan Baishya, Nilofer and Ashley Mendonca, Sunil Thomas and Anand Jain, Navroz Mahudawala and the IT team at Intrick were always there when we needed air cover and additional resources. How do we even begin to thank you?

Nachiket Mor, Vinayak Chatterjee, Sanjeev Aga, Rajiv Bajaj, Sanjoy Bhattacharyya, Sanjay Bhandarkar, Venkat Krishnan, Visty Banaji, Prakash Iyer, Rajiv Kaul, Navanit Narayan, Uday Chaturvedi, Tom Hyland, Satish Pradhan, and Anirudha Dutta—your continued guidance and intellectual support means a lot to us.

Priya Doraswamy, our literary agent, thank you for all the advice and for taking on the task to represent us globally.

There is much we owe to the team at Oxford University Press India for placing their faith in us, living up to their promise to back us up, and bringing this book to life.

On a personal note, Ramnath would like to thank Anand Krishnamoorthy, V. Balakrishnan, Chandra R. Srikanth, Karunakar Rayker, Kunal Talgeri, Manish Sharma, Srinivasan Namaji, and Professor Sudhir Bhaskar for explaining the project from different perspectives and being patient sounding boards.

Charles owes much to the prodding from his mother Lilly and brother Collins. When needed, support appeared in the form of Justina Litto and Shashikant Shetty, while Achyut Nayak played devil's advocate when it was called for. And when driven to the wall, Pearl Thomas was always there to soothe his frayed nerves.

Both of us now know why spouses are called better halves.

Sharadha N.V. put up with Ramnath's odd hours, constantly encouraging him to think from first principles and aim for simplicity.

Anna gave up on much of her own dreams and put up with Charles' cantankerous and unpredictable moods so that he may chase his dream. He will remain forever indebted to her.

N.S. Ramnath and Charles Assisi
Bengaluru and Mumbai
August 2018

Preface

IN 2009, WHEN NANDAN NILEKANI QUIT HIS JOB AT Infosys to head the government's national identity project, we were at work to build out *Forbes India*—Charles as an editor in Mumbai, Ramnath as a reporter in Bengaluru.

The news excited us *and* made us sceptical. In India, it's not common for business leaders to take up roles in government. But then, everything we knew about the government—whimsical politicians, entrenched bureaucrats, powerful lobbyists, influential networks of activists, journalists, and lawyers—suggested it is hard for an outsider to succeed. To paraphrase the legendary American investor Warren Buffett: When a manager with a reputation for getting things done enters an arena with a reputation for keeping things difficult, it is the reputation of the arena that remains intact.

Yet, when we spoke to people around, we found the general mood was one of optimism and much was being speculated upon. Many—even insiders who claimed to be in the know—thought that when the ID came to life, it would be a smartcard of some kind. India had experience with creating ID projects at scale, such as for those eligible to vote; and global entities such as Visa and Mastercard were managing hundreds of millions of smartcards. Nilekani seemed the right man for the job. For nearly 30 years, he had managed technology projects for billion-dollar businesses across the world. Now, he had to bring his experience into the government. As journalists, one question interested us the most: How does he plan to do it *at scale*, for over a billion people?

Nilekani set to work almost immediately. There was a constant flow of information on the project. We learnt it would be a biometric-based online ID, and not a smartcard. This was the first time such a technology was being issued—and at such a scale. In a little more than a year, the first ID was given out. To pull that off, Nilekani and his team of bureaucrats had built an altogether new kind of organization. It was a combustible mix of a government office and a start-up. Looking back, it's not surprising that our first story on it (which Ramnath did along with his colleague Mitu Jayashankar[I] in Bengaluru, with Charles acting as a sounding board) focused on the clash of cultures.

We followed the project with much interest as it scaled up and enrolled millions of people, before hitting a wall at 200 million. Inter-ministerial fights, bureaucratic turf wars, lobbyists of all kinds playing in the background, the rising crescendo of activists opposed to Aadhaar—all played a part. The noise may have hurt those who were at work on the project. But to us journalists, this was music. It had all the elements of a great story: interesting, complex, larger-than-life characters, conflicts, twists and turns.

These conflicts became bigger and more compelling over the years. On the political front, the National Democratic

Alliance (NDA), led by the Bharatiya Janata Party (BJP) and its allies, fought the project tooth and nail in the run-up to the general elections held in April–May 2014. One of their promises to the electorate was the dismantling of Aadhaar if elected into office. The Congress-led United Progressive Alliance (UPA) government that kick-started the project was voted out of power. On his part, Nilekani got into the electoral fray as well to represent Congress. He lost by a huge margin of over 200,000 votes.[2]

Whatever would happen next? On assuming power, instead of dismantling it, the NDA pressed the accelerator on Aadhaar, and Prime Minister Narendra Modi, who was among its most vocal opponents when in Opposition, morphed into the face of the project.

Our personal lives had changed as well. Charles took time out to start work on Founding Fuel with Indrajit Gupta, the editor of *Forbes India*, and C.S. Swaminathan, who was just done with his stint at Pearson India. Meanwhile, Ramnath went on a sabbatical to complete a fellowship at the City University of New York on entrepreneurial journalism. These new experiences, and discussions with the core team at Founding Fuel, gave us a fresh perspective to look at Aadhaar. While both of us have tracked technology for close to two decades and witnessed hype and bust cycles, such as the dotcom boom and bust, we figured there'd be a different narrative this time around.

Egged on by Indrajit and Swami in 2016, we discovered that some of the stars from Aadhaar's founding team were applying their lessons from there to build more public platforms—for example, the now famous Unified Payments Interface or UPI. It struck us that in all the hype and hoopla that surrounded the news, we hadn't paid attention to the philosophy the team had embraced.

Pramod Varma, the chief architect of Aadhaar, described it using a simple analogy: that of a Lego block. Aadhaar is a Lego block; UPI, another; eSign, one more. There are more Lego blocks built by others in the system, and can be used to build solutions. The scope of applications is limited only by the imagination.

This analogy changed the way we looked at the debate on Aadhaar. It was now all over the news and seemed focused on the downstream phenomenon—poor implementation, privacy, exclusion, fraud. We were equally intrigued by the upstream phenomenon: If Aadhaar is really just a single Lego block, what are the other pieces? Who is at work on it? Are the other pieces steeped in technology as well? Or are there institutions, processes, rules, laws … and people? Who decides what it will finally look like? Will there be one platform? Or are many platforms evolving?

Everyone at Founding Fuel was struck by the simplicity of the technology and the complexity of the whole system. This was a story that had to be explored—and a story that must be told.

It fitted in with Founding Fuel's larger mandate as well: bring out a series of books on the most compelling themes of our times that capture stories of entrepreneurship and leadership so others can learn from them.

There was no time to waste. We started having conversations with people in different ecosystems, first-hand. What struck us most was the diversity of opinions (in India, when we speak to five people, we end up getting eight opinions). It was as if they were speaking about different animals altogether. The elephant and six blind men was a metaphor that often crept into our mind.

At the same time, these interactions also helped us look beyond the binary nature of the debates around Aadhaar. Is it good or bad? Are you pro- or anti-Aadhaar? The issues were more complex. Encouraged by Arun Maira, former chairman of the Boston Consulting Group India, we turned to the tools of systems thinking—to look at second-order and unintended effects of Aadhaar.

Our discussions with these diverse sets of people were freewheeling, and moved seamlessly from technology to politics and society, and, without our realizing it, often to philosophy. What is the role of the state? What is the meaning of privacy? What is a fair way to redistribute wealth? What kind of laws do we need in a digital age? Does the acceptability of paternalism change depending on who practises it? Eventually, it struck us that Aadhaar itself has become some kind of metaphor for different people, prodding them to ask questions about power, justice, and fairness.

One of the luxuries of being journalists is this: When we face big questions, we pull out our notebooks and pens, and attempt to find answers.

Notes

1. Mitu Jayashankar and N.S. Ramnath, 'UIDAI: Inside the World's Largest Data Management Project', *Forbes India*, 29 November 2010, http://forbesindia.com/article/big-bet/uidai-inside-the-worlds-largest-data-management-project/19632/1, viewed on 30 July 2018.
2. Sharath S. Srivatsa, 'Richest Candidate Loses Heavily', *The Hindu*, 17 May 2014, https://www.thehindu.com/news/cities/richest-candidate-loses-heavily/article6017552.ece, viewed on 30 July 2018.

Prologue

The Future

ONE EVENING IN THE SUMMER OF 2017, WE MADE a day trip to a village about 60 km east of Bengaluru to talk to a group of women who had been approved bank loans for the first time in their lives. We had spent the previous few days with lawyers, technologists, policy wonks, and activists discussing Aadhaar, India's biometric identity project, through the lens of technology and public policy, digital economy and exclusion—and most importantly, privacy. Privacy was a hot topic at that time. Many were angry that the government was forcing people to link Aadhaar with bank accounts, mobile phones, and PAN cards—and took the government to the court.[1] They argued that it violated their right to privacy.

As we drove down, we were still mulling over those arguments. Presently, we found ourselves sitting on a straw mat in front of seven or eight women. We asked them what they did for a living; and with all the talk about developments in technology and rural finance being bandied about, what had really changed.

The conversation began somewhat haltingly, but eventually picked up pace.

The women, who were in their thirties and forties, shared their experience about taking loans from banks, comparing it with borrowing from micro-finance institutions (MFIs). With MFIs, they said, they had to form a group to get loans. This was not always a comfortable or comforting proposition, for there were nosy people in the group. With the help of a social enterprise—and a process that involved getting Aadhaar, submitting electronic Know Your Customer (KYC) details to banks, and undergoing training in personal finance—they were now getting loans straight from the bank. It actually enhanced their privacy—their finances were a private matter between them and their bank.

We always knew that we were dealing with a complex subject. Aadhaar was at the intersection of technology, politics, society, and business. It evoked strong reactions from almost everyone we spoke to, reminding us of David Hume's assertion: Reason is, and ought only to be the slave of the passions, and can never pretend to any other office than to serve and obey them.

We knew that the project had touched different people in different ways. Some loved it, some hated it—often from a distance that only privilege could offer. Many who were closer to the ground were starting to experience the conveniences of a digital world that Aadhaar opened up. Some struggled to adjust. And some found themselves excluded from the services guaranteed by the government.[2]

But, listening to those women that day was a novel experience. We could recognize in them a sense of liberation, a relief from the indignity of having to form a group with strangers to borrow a few thousand rupees. They didn't use the word 'privacy', but it was clear what they were talking about. That day, it struck us—more than ever—that there are no simple answers to the questions we had about Aadhaar and its effect on the country.

* * *

In one way, the story of Aadhaar is the story of the government (*sarkar*) trying to leverage technology to become more efficient and make things better for the poor. India spends billions of dollars as subsidies, and no one really has any sense of how much of it actually reaches the intended beneficiaries. Former Prime Minister Rajiv Gandhi once said that only 15 paise of every rupee spent by the government reached its target.[3] Former Planning Commission Deputy Chairman Montek Singh Ahluwalia called for 1% of all government project costs to be spent on monitoring and tracking.[4] The government hoped Aadhaar would help kick-start a process that would achieve all these. 'Aadhaar' means 'foundation' in many Indian languages, and that's what the government hoped this initiative would be: a foundation to enhance its own capacity.

For a while now, businesses have been looking for new ways to grow, especially as their goals and methods are under increased scrutiny. Top business leaders, including Unilever's Paul Polman, McKinsey's Dominic Barton, and Wipro's Azim Premji, have talked about a crisis in capitalism.[5] And across the world, business leaders have been feeling a sense of urgency to redefine the purpose of business. Interestingly, this comes at a time when the challenges to growth have turned tougher. Is it possible that innovating for the poor will achieve both purposes—make capitalism inclusive and keep businesses on the growth track?

For long, the rich have been enjoying the benefits of a digital economy. They bought their stuff online, and paid via digital cash. Few among the rich would be willing to give up their digital tools—smartphones, credit cards, social networking sites. Big businesses and start-ups are vying to innovate for them. The poor are mostly deprived of these benefits.

Transaction costs are high, which makes low-ticket-size products and services unviable. Many businesses avoid this segment. For example, mutual funds are not interested in those who have just Rs 500 or Rs 1,000 to invest, because servicing a low-value customer would eat into their profits.

By building digital public infrastructure, such as Aadhaar, the government hoped to bring transaction costs down, just like how building physical infrastructure such as roads and railways can bring travel time down and give easier access to places that few go to. Thus, the story of Aadhaar is also the story of businesses (a key player in the *bazaar*) trying to leverage the public digital infrastructure to serve more customers, and serve itself.

As the government and businesses rush to build solutions on top of Aadhaar, NGOs, social enterprises, activists, academia, and the media (other key players in the bazaar, the marketplace of ideas, resources, action, and influence) have been playing an interesting role too. For one, they have been highlighting the unintended consequences and inherent risks of the project: exclusion, misuse by private parties and government, faulty implementation.

Living up to the description given by economist Amartya Sen, 'argumentative Indians' have been discussing Aadhaar threadbare. So far, there has been more sound than light. The shrill noise of polarized debate has left the intended beneficiaries, the society (*samaj*) confused. It's as if there can only be two options—either Aadhaar is good, or it is bad; either you can be against it, or for it. You form groups based on these answers. You become intensely loyal to your group and start looking at others who don't agree with you as enemies. Such polarized arguments will not dispel this confusion; only real benefits to samaj will. That has just begun to happen. For example, 600 million Aadhaar biometric authentications take place every month.[6] Indians transferred over Rs 450 billion on India Stack's UPI in over 235 million transactions in July 2018 alone.[7] This will grow, as more innovations happen on Aadhaar and India Stack (the applications built around it)—but people are already taking note.

* * *

In reporting and writing the story, we tried to be loyal only to the truth. And that approach helped us see that the story of Aadhaar holds important lessons for everyone.

For business leaders, Aadhaar and India Stack open up a whole set of new opportunities. Some of the smartest entrepreneurs and businessmen are looking at ways to leverage these technologies to build innovative products and redesign their business models.

For those who are thinking of making a transition to government or to the social sector, Nandan Nilekani's experience in UIDAI and the way he navigated the system can offer many valuable lessons.

For policymakers, there are lessons on combining technology with governance. Contrary to popular perception, Aadhaar is not a technology solution, but a platform, a Lego block that can help in building technology solutions. Like everything else, it comes with advantages and disadvantages.

For technologists, Aadhaar and India Stack offer an excellent case study of building platforms at scale—digital public infrastructure for a billion people and more.

For social entrepreneurs, there is the whole new world of Societal Platforms—conceptualized from the learnings from building Aadhaar and India Stack.

For activists who want to develop a broader perspective and sharpen their systems-thinking skills, the story of Aadhaar is a story of interconnections and second- and third-order impacts.

* * *

Aadhaar is very much a developing story. There are those in the government who understand the import of this project on India and the potential it holds to reimagine capitalism across the world. But it is still a work in progress, and countries from the developed and developing world are looking at the potential outcomes with much interest.

To cite but a few instances, the US is looking at open authentication API (Application Programming Interface), a feature missing in its social security number system; its agencies are also studying India Stack, especially UPI, as a financial infrastructure for the country. Sri Lanka is looking to launch an Aadhaar-like initiative in the country.[8] Bangladesh has sought help from iSPIRT, a think tank for the Indian software products industry, to see what lessons it can learn from Aadhaar to improve its own identity programme. Russia, Morocco, Algeria, and Tunisia have shown interest in the Aadhaar model.[9] In 2017, a delegation from the UK was in Bengaluru to study both Aadhaar and India Stack, and there are talks that it might launch a national identity project (after having dismantled its earlier initiative in 2010). There are at least 20 countries that have shown interest in India's digital infrastructure.[10]

But even as the system evolves in India, it has to not only fight the pulls and pressures within the bureaucratic and political system, but also function within a vibrant democracy. As it goes about doing this, there are voices of all kinds—from politicians, bureaucrats, online activists, social workers, businesses, lawyers, journalists. Each has their own version of the story.

If your quest as a reader is to find harmony in this narrative, be assured there is much chaos in the story that follows. That is what makes this an incredibly compelling narrative.

Even as we were at work to put the final touches to this manuscript, a five-judge bench of the Supreme Court listened to arguments on Aadhaar for over 38 days, spread over nearly four months—the second longest hearing in India's history.

A group of petitioners wanted Aadhaar to be scrapped (for reasons we will explore in the book). The government and a few other institutions defended it. The arguments have spilled outside the court and into drawing rooms. It suggests how contentious this subject has become.

That is why there is no better time to start reading this story than now.

A Lego Block View of Innovation

As you join us to explore the genesis and evolution of Aadhaar and the other technologies described in the narrative that follow, we urge you to deploy a metaphor. Imagine each of these technologies as Lego blocks—the interlocking bricks invented by Ole Kirk Christiansen, Danish inventor and founder of The Lego Group. These pieces can be used to build objects of all shapes and sizes: cars, aeroplanes, submarines, buildings, or, for that matter, superheroes and villains. This is best left to the imagination of those using them. Aadhaar is just one brick of a Lego block. This story is about how a team of experts came to build it, and how it's being used by sarkar, samaj, and bazaar.

In Chapter 1, *The Hare and the Tortoise*, the focus is on the organization that Nandan Nilekani and his team created to roll out Aadhaar. Described by many as a start-up within the sarkar, it had the characteristics of both. Those characteristics gave the team unbeatable advantages as well as far-reaching disadvantages. The chapter explores them, while narrating the story of how Nilekani built the team.

In Chapter 2, *The Art of War*, we look at how Nilekani took on the forces that felt threatened by the project—for good reasons and bad. As an outsider in the country's capital, known for political intrigue and backstabbing, Nilekani had to dig deep into his reserves of strategies, mental models, and tools to take them on.

Chapter 3, *The Platform Paradox*, explores how Aadhaar is being used to build solutions on the ground. It was used in a variety of schemes—mostly to deliver government benefits. Have they worked? That depends on what the agencies did with the Lego block.

If Aadhaar is one piece in a Lego construction set, Chapter 4, *When a Butterfly Flaps Its Wings*, tells the story of how other pieces came to be built. These include UPI, eSign, eKYC, DigiLocker, and Data Empowerment and Protection Architecture (DEPA), collectively known as India Stack. Together with other infrastructure that the country is building, primarily the Goods and Services Tax Network (GSTN), these pieces could have far-reaching effects.

In Chapter 5, *Insurgents, Incumbents, Pioneers, and Leaders*, we explore how these Lego blocks—Aadhaar, India Stack, and other infrastructure—are impacting the world of business. There are interesting lessons from start-ups, large organizations, and global leaders.

In Chapter 6, *The Naysayers*, we look at the profiles of people who are opposed to Aadhaar. Some of them worry about the way the Lego blocks have been used to build solutions, and some want the Lego blocks to be destroyed. Why ought it attract such fierce emotions? Our research suggests there are 50 concerns people have on their minds.

In Chapter 7, *India and the World*, we look at how the fundamental idea of Aadhaar has evolved into Societal Platforms. These are the digital Lego blocks that can be used for social transformation, for example, by educating millions of students. It also explores how digital public infrastructure can be used to build India's soft power.

Notes

1. Apurva Vishwanath, 'Aadhaar Linkage with PAN Mandatory, Rules Supreme Court', *Mint*, 10 June 2017, https://www.livemint.com/Politics/Nd21HZXkXbN2abOki9GPYJ/Supreme-Court-verdict-on-Aadhaar-linkage-with-tax-returns-P.html, viewed on 26 June 2018.
2. Jean Drèze, 'Why Linking Aadhaar to PDS Threatens to Disrupt Food Security', *DailyO*, 24 October 2017, https://www.dailyo.in/politics/pds-biometric-aadhaar-card-public-distribution-system-bpl-apl/story/1/20208.html, viewed on 26 June 2018.

3. PTI, 'Rajiv Gandhi's Popular 15 Paise Remark Finds Mention in Supreme Court Verdict', *Indian Express*, 9 June 2017, http://indianexpress.com/article/india/rajiv-gandhis-popular-15paise-remark-finds-mention-in-sc-verdict-4696740/, viewed on 26 June 2018.
4. Rediff.com, 'PM Sets Up Monitoring Group to Track Big Projects', 13 June 2013, http://www.rediff.com/business/report/pm-sets-up-monitoring-group-to-track-big-projects/20130613.htm, viewed on 26 June 2018.
5. Daniel Gross, 'How Paulson Became the New Face of Capitalism', *Newsweek*, 19 August 2008, http://www.newsweek.com/how-paulson-became-new-face-capitalism-88637, viewed on 26 June 2018.
6. Aadhaar Dashboard, UIDAI website, https://uidai.gov.in/aadhaar_dashboard/auth_trend.php?auth_id=biofingure, viewed on 6 August 2018.
7. NPCI, 'UPI Product Statistics', https://www.npci.org.in/product-statistics/upi-product-statistics, viewed on 6 August 2018.
8. Meera Sreenivasan, 'Sri Lanka Is Keen to Introduce an Aadhaar-Like Initiative', *The Hindu*, 21 December 2017, http://www.thehindu.com/news/international/sri-lanka-is-keen-to-introduce-an-aadhaar-like-initiative/article22099494.ece, viewed on 26 June 2018.
9. Amrit Raj and Upasana Jain, 'Aadhaar Goes Global, Finds Takers in Russia and Africa', *Mint*, 9 July 2016, https://www.livemint.com/Politics/UEQ908E08RiaAaNNMyLbEK/Aadhaar-goes-global-finds-takers-in-Russia-and-Africa.html, viewed on 26 June 2018.
10. Jayadevan P.K., 'India's Latest Export: 20 Countries Interested in Aadhaar, India Stack', *Factor Daily*, 10 January 2018, https://factordaily.com/aadhaar-india-stack-export/, viewed on 26 June 2018.

I

The Hare and the Tortoise

ONE FRIDAY EVENING IN AUGUST 2009, THE AIR IN the sprawling campus of India's second largest IT company was thick with excitement—there was a mix of pride and anticipation. Nandan Nilekani, the company's chairman, was to deliver his farewell address. He was quitting Infosys to join the government on a critically important assignment. The employees were keen to know what he would say and get a sense of what he was feeling.

Nilekani, then 55, was known to be unsentimental. He had a cerebral, even clinical, approach to life. But that day he seemed to have let himself be guided by his heart. 'I am generally very articulate, but this is not the day nor the place where I can be articulate. I've been wrapped up in Infosys for 28 years. My only identity is Infosys. I will be going to lead a programme to give identity to every Indian. But today I am losing my identity,' he said.

If there were gods who looked over what people say in their farewell speeches, they would have laughed. For, nine years after Nilekani gave that speech, he was back in Infosys as its chairman. And far from losing his identity, he stamped an identity on a project that so indelibly became his.

On the face of it, identity might not appear to be important. But India was facing an 'identity crisis'. The government, which was spending billions of dollars on subsidies, was not sure it was reaching its target. Rajiv Gandhi, India's sixth prime minister, had famously observed that only 15 paise of every rupee spent by the government actually reached its target. There were leakages, and chances were extremely high that millions of poor who needed government subsidy were not getting it, because the government had no reliable way to identify them.

(However, the question was not just about efficiency. A good identity system can give value to people. 'Unique identification of each citizen also ensures a basic right—the right to "an acknowledged existence" in the country, without which much of nation's poor can be nameless and ignored, and governments can draw a veil over large scale poverty and destitution,' Nilekani had written in his book *Imagining India*, published a year earlier.)

If the state has a way to identify people—as individuals—it can give them their entitlements irrespective of where they are. This means a person can migrate from Bihar to Tamil Nadu for a few months, and all they would have to do is prove their identity to the government to get their rightful benefits.

A well-designed identity platform can even spur innovation. GPS (Global Positioning System) and TCP/IP (Transmission Control Protocol/Internet Protocol) were built as platforms and, eventually, government and businesses started building solutions above them, creating new kinds of industries.

Identity is not just an Indian problem. According to the World Bank, gobally, one in seven people are unable to prove their identity, most of them under the age of 18.[1] Without a reliable identity system, the World Bank contends that countries struggle to govern effectively and make efficient use of limited resources. It is not just a problem for sovereign states, but also for multilateral agencies, such as the United Nations High Commissioner for Refugees, which have not been able to find a way to give identities to thousands of people driven out of their countries because of violence and war.

Providing a legal identity to everyone by 2030 is among the sustainable development goal targets.[2]

If Nilekani succeeded in providing an identity to everyone in India, it could be a template for other countries and development agencies. After all, India has a billion-plus citizens who are a hugely diverse set—from globetrotters to those who have never travelled beyond a 5-km radius. If one can solve it for India, one can solve it anywhere.

But it was not going to be an easy task. Government projects are faced with cost and time overruns even in things that have been done over and over again, like building roads or dams. Previous attempts by the government to use technology to give identity indicated that there was a high chance of failure. In 2003, when Atal Bihari Vajpayee was India's prime minister, the government launched a pilot to give identity cards to citizens. The first set of identity cards were given to residents of a locality in Delhi—four years later.[3]

Nilekani was clearly stepping out of his comfort zone. He had spent nearly 30 years in a single organization. He was working with people like N.R. Narayana Murthy with whom he had differences in terms of style, but not in terms of substance or values. In many ways the Infosys co-founders were like a family that had stuck together through the ups and downs.

Working in government was going to be different, and Nilekani was fully aware of that. In his farewell speech, he said,

'I am leaving an organized world. Here, standing at the top of an abyss, even if I were to fall, I may find water. But, in my new role, I'm supposed to work with 600 government departments knowing fully well that no two government departments get along with one other.'

'We think it's all easy. But the world is not like that. One of my favourite books is *Father, Son & Co* by Thomas Watson. You might have thought IBM was all professional. You read the book and you see you have these two brothers, Tom and Dick, competing for the job. So, the father divides the company into two for his two sons. One is IBM USA and the other is IBM World Trade Corporation. Tom then marginalizes his brother, and takes control of the company. When he becomes the CEO, his sister sells all the shares in IBM because she doesn't have confidence in Tom. You realize business is complex. Life is messy,' Nilekani told us.

Bengaluru Meets Delhi

No one really expected Nilekani's move to government to be easy. In IT campuses across the country there were jokes about Nilekani, a quintessential private sector guy from Bengaluru, struggling his way around New Delhi, the bastion of babudom.

In one joke, a bureaucrat asks Nilekani how he finds Delhi. 'Delhi?' Nilekani immediately drags him to a room and gives a presentation on the city, what its pain points are, and how they can be fixed. The said bureaucrat requests for a posting back to his state.

In another, Nilekani, frustrated about how slow things run in Delhi, complains to Montek Singh Ahluwalia, the deputy chairman of the Planning Commission. Ahluwalia thinks for a while and says he should take this to the prime minister. Manmohan Singh listens to Nilekani and promises to get back soon. A month later, Singh calls up Nilekani:

'We are happy to say we have found a solution for your problem, Nandan*ji*.'
'That's fantastic. Thank you.'
'Yes, we have decided to form a committee to study speed in bureaucracy. Would you like to head it?'
'But, Mr Prime Minister …'
'Don't hesitate, Nandan*ji*. There is no hurry. You can take five years to submit the report.'

Though a statement on the way things ran in Delhi, these jokes reflected the popular opinion—that Nilekani was taking on a tough job, and that he was more likely to fail than succeed.

To succeed anywhere is tough. To succeed in this particular endeavour seemed even more so, to anyone who knew the system and how different the world Nilekani came from was from the one he was getting into. 'My first reaction was that it [the project] would not survive the political economy,' Manish Sabharwal, a Wharton MBA who founded temp staffing firm Teamlease, told us. Sabharwal has more than a passing interest in public policy and has worked closely with the state in the areas of employment and skill development. 'To me, the implication of what he set out to do became obvious. I thought the status-quo-ists would have made the same connections, and that they would recoil—the antibiotic reaction of the Indian state. People don't cut the tree they are sitting on. And Nilekani was going from a place that asks "how do we do it?" to a place where anybody can say "no" and nobody can say "yes".'

In some ways, Delhi wanted Bengaluru. Politicians, bureaucrats and policy experts were thinking about a universal, trusted, national ID—for either security reasons or efficiency reasons. L.K. Advani, who was deputy prime minister from 2002 to 2004 when the Bharatiya Janata Party (BJP)-led National Democratic Alliance (NDA) was in power (1998–2004), was keen on rolling out a national ID project. In his 2008 report on financial sector reforms,[4] economist Raghuram Rajan (who would later serve as the government's chief economic advisor and governor of the central bank, Reserve Bank of India) highlighted the need for an ID as a basis for finance. Economist Arvind Virmani, the chief economic advisor between 2007 and 2009, had worked on a detailed plan. One of the last big projects of management guru C.K. Prahalad was to present a letter to Prime Minister Manmohan Singh, head of the Congress-led United Progressive Alliance (UPA) government, persuading him to launch such a project.

These initiatives had cumulatively resulted in the government setting up the Unique Identification Authority of India (UIDAI) in 2008. The need for such an institution was felt more deeply because the UPA government had stepped up on the targeted subsidies and benefits programme, including the launch of an employment guarantee law and a food security law. One of the boxes they wanted to tick was to be able to identify the beneficiaries. After the Congress-led UPA coalition's spectacular win in the 2009 general elections for a second term, the Congress party felt confident enough to bring outsiders into the government, for it no longer had to depend on the Communist party to claim the office.

John Kingdon, a professor of political science at the University of Michigan, has argued that for a window of opportunity to open, three streams must come together: the problem stream—the recognition that there is a problem; the policy stream—the availability of solutions; and the politics stream—politicians willing and able to make the policy change.[5] A number of things can open this window—including public mood and the collective will of a few politicians. But a number of factors can quickly close this window too.

In 2009, these three streams came together in India. Identity was being seen as a hurdle in the delivery of subsidies and other targeted public goods; there were a number of options to solve the problem; and there was political will—driven by Congress party president Sonia Gandhi, probably the most powerful politician of that time. What was needed was someone who would be able to make full use of the window of opportunity—and deliver.

They turned to Nilekani.

A Corner of a Foreign Field

Nilekani had prior experience working with the government. From 2005 till 2008, he was a member of the Knowledge Commission, a group of experts that advised the prime minister on education, science, technology, and governance. In 2003, he had co-founded a non-profit called the eGovernment Foundation, along with technologist Srikanth Nadhamuni, which aimed at solving urban governance problems using technology. In 1999, Karnataka chief minister S.M. Krishna had invited him to chair the Bangalore Agenda Task Force, a partnership between citizens, businesses, and government agencies such as police and municipal corporations.

Nilekani's mental model of government and bureaucracy was not shaped by his experience alone. One of his most valued assets was his Rolodex. Whenever he wanted clarity on any subject while writing *Imagining India*, he would pull out his Rolodex, contact some of the best people working in the field, and get a masterclass. He followed pretty much the same formula before he took up the job at UIDAI. He spoke to his friend K.P. Krishnan, a cadre Indian Administrative Service (IAS) officer—and also to others such as T. Koshy, executive director at National Security Depository Ltd, and Sriram Raghavan, founder of Comat Technologies, an e-governance service provider that had built a public distribution system (PDS) for the Karnataka government.

From all these conversations he knew that it was possible to build a digital identity infrastructure for a billion people, but he also knew it was going to be difficult. If he tried to do it the traditional (read government) way, it was not going to work. He had to do it in a completely new way.

And for that he needed space—physical, bureaucratic, political, and even intellectual. So, before he agreed to do this, he ensured that he would have the rank of a cabinet minister and while making the announcement the prime minister would explicitly say he had invited Nilekani to take up that position.

These things matter in government, a senior bureaucrat based in Delhi told us. 'Contrast this to Natgrid, which was another government project that came up around the same time. I have no doubt that Captain Raghu Raman, who headed the project, is capable. But he came in with the rank of a secretary, and not of a cabinet minister. Which meant that he had to go through the whole hierarchy to get anything done. That slowed him down. Nilekani, on the other hand, had the option to pick up the phone and call the prime minister if he wanted to. Within the system, what matters is not whether he *would* do it; what matters is he *could* do it. That's how power works.'

That didn't mean Nilekani could have his way. The rank, the perception, the halo simply gave him some muscle to push back a system endowed with, as Montek Singh Ahluwalia told us, too much negative power. 'A bureaucrat can be questioned for buying a ream of paper without proper approval—which might have cost the government all of Rs 200,' says P. Ananthakrishnan, a former bureaucrat and author of *Muddy River*, an account of his attempts to negotiate with ULFA, a separatist outfit based in Assam, to release an abducted engineer. 'The same bureaucrat,' he says, 'can stall a mega project on some technical issue—costing hundreds of crores of loss for the exchequer and lost jobs. Not only would he not be questioned, he would be praised for doing his job well.'

The World Is What It Is

The real world is complex and dynamic. Everything is interconnected, and you can't do just one thing—because any action can set a series of other events in motion. Not all of them can be anticipated. Often they are counterintuitive—what seems like a perfectly rational decision today might come back to bite you because circumstances have changed. And the real world reacts in unexpected ways. People resist, they adapt, and then go back to the old equilibrium.

This is true even in relatively stable organizations. Nilekani was trying to build something altogether new. Building a place where, to hark back to an Aesop's fable, an agile hare and a methodical tortoise can work together. In the fable, a tortoise—constantly mocked by a hare for being slow—challenges it to a race. The race begins. The hare is off a stunning start and halfway through, it is so confident that it decides to take a nap. A Usain Bolt could well have pulled it off, but the hare, when it wakes up, realizes that the tortoise, going step by step, had crossed the finish line.

Nilekani was flipping the story. Why make them compete? Why not let them cooperate?

In creating this new way of working with the government, Nilekani's argument went like this: a government is good at scale, it operates across the country; a start-up is good at speed, it can bring a product from its design stage to the market in a matter of months.

A government has stability, and it can run at the same pace for years; a start-up has agility, and often chants the mantra, 'fail fast, fail often'. In fact, thought leader Steve Blank defines a start-up as 'a temporary organisation designed to search for a repeatable and scalable business model'.[6] Governments are good at staffing their rolls with generalists—administrators who can handle every type of project. 'If you are in bureaucracy long enough, you get a broad perspective,' Ananthakrishnan says. For instance, an IAS officer recently shared the roles he played in his district: managing director of a power distribution company, member secretary of pollution control committee, director of social welfare, secretary of labour and employment, and commissioner of tribal welfare, among others. A start-up, on the other hand, is good at hiring domain experts who can solve a specific type of problem fast.

Governments are good at designing incentives that are based on processes rather than outcome. For bureaucrats, as economist Thomas Sowell said, 'procedure is everything and outcomes are nothing'.[7] That's also true of many large organizations operating in a relatively stable industry. In a start-up, the incentives are shaped by the market. Even where traditional metrics such as sales and profits don't apply, start-ups look for signals from the market to see if they are delivering value, and where they can do better.

Nilekani, looking at an ambitious goal and aggressive deadlines, wanted to merge the scale of the government with the speed of a start-up. He wanted a government department's stability and a start-up's agility, the bird's eye view of a generalist and the deep knowledge of a domain expert. And at the same time, find a way to leverage the power of the market while keeping in line with the exacting processes dictated by the rules of bureaucracy—processes necessary for a project that would impact a billion people.

Nilekani knew he would not be able to do it alone. He needed good partners, both in the government and from the private sector—partners who could work with him on equal terms. To execute a project of this scale and complexity, he would also need expertise from a range of sectors, and therefore he would have to embrace diversity. At the same time, he would have to ensure that UIDAI didn't become so big that it closed itself to market-based incentives, but was big enough to ensure due processes were followed.

However, life often behaves like a vegetable vendor who adds one or two rotten tomatoes along with good ones. Often, things come to us as bundles—good and bad, blessings and curses. Strong-willed people come with their own vision of how the world should be. Diversity often leads to innovation, but can also bring conflict. And as every person trying to cut flab knows, muscles go in the process too often.

The Go-To People

In the world of business, politics, and society, there is a myth of the lone genius. A popular advertisement by One Earth, an environmental group, extols 'the power of one' by showing clippings of Mahatma Gandhi (wanted to raise consciousness without raising his voice) and Mother Teresa (travelled the world to give hope to those who had none).[8] Walter Isaacson, who has written widely acclaimed biographies of Steve Jobs and Albert Einstein, points out that if you search the phrase 'the man who invented' on Amazon, you get 1,860 book results.[9]

Yet, the history of technology and business shows that no one can invent anything all on their own. Inventors not only stand on the shoulders of giants before them, but also have several hands to support them.

In *Powers of Two*, Joshua Wolf Shenk[10] wrote:

> For centuries, the myth of the lone genius has towered over us like a colossus. The idea that new, beautiful, world-changing things come from within great minds is now so common that we don't even consider it an idea. These bronze statues have come to seem like old-growth trees—monuments to modern thinking that we mistake for part of the natural world.

Wolf Shenk chronicled the lessons he learnt from studying creative pairs, not only the well-known ones such as John Lennon and Paul McCartney of The Beatles; Steve Jobs and Steve Wozniak of Apple; Marie and Pierre Curie, who discovered radioactivity—but also partnerships like those of Vincent Van Gogh and his brother Theo, and even Tiger Woods and his caddie Steve Williams.

The dyad, Wolf Shenk wrote,

> ... is the most fluid and flexible of relationships. Two people can basically make their own society on the go. When even one more person is added to the mix, the situation becomes more stable, but this stability may stifle creativity, as roles and power positions harden. Three legs make a table stand in place. Two legs are made for walking or running (or jumping or falling).

The last thing Nilekani wanted to do was to stand in place.

Towards the end of 2017, a good four years after he had quit UIDAI, Nilekani seemed to be juggling too many balls. Besides being the chairman of Infosys, he was running a venture capital fund called Fundamentum, an education platform called EkStep, an NGO called eGovernments Foundation, and on top of it, he was travelling the world to convince influencers that the time was ripe for what he called Societal Platforms. We asked how he gets time for all this.

'I don't watch cricket. I am not into Twitter—I only use it as a broadcast medium—and I am obsessive about not wasting time. Every minute is important to me,' Nilekani told us.

'The projects I take up, I don't do it alone. I work with strong people. They are my go-to persons. At Infosys, it's the CEO [Salil Parekh]. At EkStep, it's [co-founder] Shankar Maruwada. With eGovernance, I go to Viraj Tyagi [CEO of eGovernments Foundation]. For technology architecture, it is Pramod Varma [chief architect of Aadhaar and EkStep].'

'I get involved in some of the projects because I happen to meet someone who is good at it. For example, Sanjay Purohit who used to work with me at Infosys had quit his job. We met and we thought we could work together. He came up with this whole idea of Societal Platforms. He is now my go-to guy for that.'

When Nilekani was working on Aadhaar, he was obsessed with it, with little time for anything else. Instinctively, he was following the same approach. It began with Ram Sevak Sharma, director general of UIDAI, who essentially ran the organization; but it was not limited to him. For financial processes, Nilekani depended on K. Ganga. For his meetings that had anything to do with finance and banking, he was almost always accompanied by Ashok Pal Singh. Similarly, for technology, his go-to guy within the system was Srikanth Nadhamuni, in Bengaluru. And for marketing, he had Shankar Maruwada.

The Guardians of the Welfare State

Nilekani, from his work with the government earlier, knew that there was no shortage of talent within the bureaucracy. The challenge was to identify them and get them to work for the project, pulling them off their own career path, while navigating the bureaucratic maze. He wanted bureaucrats who had an entrepreneurial mindset, who knew how things worked within the government, who could work with a diverse set of people—without for a moment forgetting that the project was ultimately to empower people and not the government itself.

Ram Sevak Sharma, by all accounts, is a typical bureaucrat. If you happen to see him in a neutral setting, say, in a restaurant, and a friend asks you to guess his profession, you are likely to say 'bureaucrat'. Sharma has a mop of white hair, walks around with confidence, and has a habit of talking about his meetings lined up for the day in a manner that nudges people to use the time with him to talk about important things.

Sharma, who studied mathematics, has a fascination for computers. He likes to use computers to solve practical problems. In 1985–86, he bought one of the earliest computers, and created a database of lost weapons. He once wrote a program for a friend—an accountant—that came to be used by stock market traders. And when he was 45, he decided to do a masters in computers, and called a professor at the University of California, Riverside, to ask if he could study under him. The professor was impressed and called him over.

When Nilekani was appointed chairman of UIDAI, he consulted his IAS friend K.P. Krishnan to build the core team. In the government, no one gets that luxury. But, it was Nilekani's early days, newspapers were raving about his move, and for the government it was considered a coup. That gave him some leeway to choose his own team.

The very first person Nilekani met was Sharma. Sharma's bureaucratic colleagues say it must have been a no-brainer for him. When Nilekani approached him, Sharma was in Jharkhand. For bureaucrats, working in Delhi was considered a step up. The project was about technology, and Sharma had a fascination for it. When Nilekani asked him if he would join him, he answered immediately. He told Nilekani, 'You have taken a real risk taking up this project. I will of course join you.'

Nilekani was happy as well. Sharma understood technology, its application in governance, and knew his way around bureaucracy. Sharma's colleagues called him *dabang*—audacious—for the way he went about his job. Once, a top minister was unhappy with Sharma for organizing a meeting of some officers, and hauled him up. 'Mr Secretary, with whose permission did you organize the meeting?' the minister is said to have asked. Without missing a beat, Sharma replied, 'Mr Minister, whose permission do I need to do my job?'

As the work got intense, Sharma's knowledge and attitude was of immense value to UIDAI. Things often had to happen yesterday. There were swarms of people representing tech companies, the social sector, and businesses knocking on their doors to see how they could help and how they could benefit. It needed someone like Sharma who was sure of the bureaucracy's place in the world and his own place in the government to deal with them.

'You will not believe the number of lobbyists and salespersons that walk the corridors of any government organization. To get a government contract not only means money today, but also money tomorrow. Government projects often mean vendor lock-in, and that also slows down the rate of improvement. Sharma knew the game very well. The lobbyists and consultants realized very soon that they cannot BS with him,' one of his colleagues told us.

Sharma knew technology well enough to read and understand the codes written by the tech team. 'Once we got a mail from Sharma regarding a module at 4 a.m. Reading the mail, it was evident that he had gone through the codes line by line,' an engineer who worked with Mindtree, which was developing the enrolment software for UIDAI, said.

One of the key things that Sharma had to get right was the dynamics between himself and Nilekani. With his previous bosses, it had been easy enough. 'The relationship was that of one between a cabinet minister and secretary. Nandan had the cabinet rank. I was in effect the secretary,' Sharma said. However, a project like this had never been done before, and that involved a lot of discussions and deliberations. In that there were a lot of pulls and pressures even between Nilekani and Sharma.

For his part, Sharma understood the need for both quality and speed. 'There is this tendency to rush through a project and then come back and fix things. I had seen too many projects that go like that. So, this had to be thought through very well. There was also a need for speed. This cannot go on for ten years, because technology is changing fast. We had to balance quality and speed,' he said.

Ultimately, it was clear to everyone that it was a government project, and the lateral entries from the private sector, the volunteers, and the interns were there to support the government to get this done. They were there for their expertise. Irrespective of what their positions were in the world outside, irrespective of their levels of expertise, each one of them was reporting to a deputy director. 'You can't have free radicals going around,' Sharma said at that time. He held informal meetings with all the six senior deputy directors every Tuesday to assess how things were going.

Many who looked at the project from outside—analysts, consultants, journalists—saw it as being driven by the lateral entries, by the private sector. Sharma and senior bureaucrats let that impression be, in part because as government servants they were supposed to be invisible. But they also saw that as an expression of the ownership they felt about the project. Sometimes, some of them took it a bit too far. A bureaucrat narrated an anecdote about Sharma attending an event in London to speak about UIDAI. Seated among the audience, he saw one of his colleagues from the private sector—oblivious to the fact that his boss was sitting in the audience and was slated to speak next—introduce himself as the person running the project. Sharma was amused. 'My problem,' Sharma told his colleagues later, 'was how do I introduce myself after that.

I didn't want to embarrass him. So, when I went on stage, I merely said I worked in UIDAI.'

This gentle treatment was of course not reserved for everyone. When he was in UIDAI, many saw him as a person living up to his reputation of being *dabang*. The partners who worked with UIDAI at that time felt he was unsympathetic to their problems, and was demanding everything to be done yesterday.

But Sharma was also a dyed-in-the-wool bureaucrat in an organization that was swarming with private sector people. He was increasingly seen as a person who represented the interests of bureaucrats rather than those from the private sector. 'Bureaucrats, in general, are turf conscious, and in-group loyalties ride high among them. He did what any bureaucrat would have done. He let bureaucrats capture UIDAI, and slowly crowded out people from the private sector. Today, it's like any other government department,' a bureaucrat told us.

Often in some of the most productive partnerships, one partner remains hidden from the public glare. They do the hard work, but behind the scenes. In *Powers of Two*, Wolf Shenk[II] talks about Gandhi's dynamics with Mahadev Desai, his secretary:

> Before Gandhi rose each morning, Desai laid out the plans for the day. After Gandhi retired, Desai made notes for the movement's official record. According to Desai's son, Gandhi would often examine Desai's texts and make only a single change: at the end, he would cross out the initials M.D. (Mahadev Desai) and replace them with M.K.G. (Mohandas K. Gandhi).

Nilekani was not exactly looking for a Mahadev Desai, but he wanted someone who would guide him personally through the bureaucratic and political maze. He knew bureaucracy is a different world, and can tire out even the most energetic outsiders, making them walk from one office to another, or bury them under a pile of files. The story of Shashi Tharoor, who joined the government as an outsider at around the same time, illustrates the point.

Tharoor had a distinguished career in the United Nations (UN). At one point it looked as if he would become the secretary general of UN, a position no Indian had attained before. However, he didn't garner enough support. He returned to India and joined the Congress party, stood for elections at Thiruvananthapuram, and got a ministerial post.

His honeymoon period didn't last long. Even as the Congress party called for austerity (Sonia Gandhi, the party's president, travelled in economy class, a couple of ministers cancelled their foreign trips), it emerged that Tharoor was staying in a five-star hotel. When someone asked if he would fly in economy class, he couldn't resist replying, 'Absolutely, in cattle class out of solidarity with all our holy cows.'[12] It was a funny tweet, but didn't go down well with the party leaders. He was rebuked.

Nilekani had never let his brilliance outshine his common sense, and he knew he needed someone who understood New Delhi, an insider who could tell him what works and what doesn't. Again, K.P. Krishnan referred him to a National Law School gold medallist and a Karnataka cadre officer, M.S. Srikar. Srikar, who was then undergoing his mandatory mid-career training at Mussoorie, attended an interview with Nilekani and joined his team.

Srikar knew the bureaucratic and political maze and would advise Nilekani on where to press and where to hold back. As almost everyone else in the early days, when something had to be done, he rolled up his sleeves to work on it. The first version of UIDAI's website was in fact built overnight by Srikar along with his colleague, K. Ganga. Srikar, his wife, and their child landed at Ganga's home one evening, and worked on the website with the basic information about the project.

Around the same time, Nilekani also reached out to Ganga, a highly regarded officer. Ganga was serving as accountant general at Shimla. (Shimla was the summer capital of British India and retained much of the grandeur and pomp that marked the British Raj—even in terms of the administration). Ganga made a trip to Delhi and met Nilekani. It was essentially an interview for the role of chief financial officer of UIDAI.

Ganga was aware that many saw an accountant as merely a person who would see if a proposal checked all the boxes and either approved or sent it back. However, she was not just another bean counter. She understood technology in the way pragmatic people understood technology. She knew the impact it can have, the resistance it could face, and above all, the eccentricities of techies.

Back in the early 1990s, when she was deputy accountant general in Chennai, Ganga had led the computerization of provident fund accounts. 'It was exciting. We had one engineer who liked to work only during the nights. He would walk in with a two-in-one with MGR [Tamil cinema superstar M.G. Ramachandran, who also served as Tamil Nadu's chief minister in the late 1970s and 1980s] audio cassettes, play them loudly, open multiple machines, and work on them almost in parallel.' The machines didn't always connect well, and her co-workers sometimes saw Ganga going around fixing the cables, with a couple of cables hanging around her neck.

Even back then, Ganga could see how transformative it would be. The computerization project in Chennai was just about one department. A unique identity programme was going to be huge. She knew some of the pieces that needed to work for it to succeed, and some of them were not just about finance.

When Nilekani asked Ganga if she would join the team in finance, Ganga had one request: that she be given an additional role in the operations. She wanted to be part of the team that created the project ground up. Nilekani, who had not yet experienced the system in full force, readily agreed.

Three-and-a-half years later, this decision would hit back at Nilekani—when UIDAI was fighting battles not only with the bureaucrats in the Home Ministry, but also with those in the Planning Commission, which housed UIDAI. 'I was told there was a conflict in my doing both finance and operations. There was none. My role in operations was more about contributing to what fields a form should have, for example. However, the intention was not to rule out conflicts, but to hurt UIDAI. So, I was given a choice to either be in finance or in operations. When I joined UIDAI, Nilekani asked me to give UIDAI five years of my life. But I had to go after three and a half. I did not keep my promise. But neither did Nilekani,' Ganga told us. For Nilekani, at that point, there were bigger battles to fight. He was advised by his well-wishers to let this go. And Nilekani reluctantly did, one bureaucrat who was familiar with those fights told us.

Not everyone waited for a call from Nilekani. In 2009, Ashok Pal Singh was driving technology at the postal department. If there was one department in the government that had a close connection with Aadhaar, it was India's Department of Post, which was in the business of identifying people. And in a country of this size and complexity, it was doing its job amazingly well. Generations of people depended on its services—implicitly believing that a postcard (which for a long time cost just 15 paisa) or an envelope dropped into the red boxes dotting the country would reach the intended person sooner or later. It had also fine-tuned the transfer of money to the farthest corners of India. It had played a major role in financial inclusion by setting up post office savings banks. India's postal system was much more than the hugely complex systems and processes that took a letter from point A to point B with accuracy. The village postmen often play a major role in not only delivering letters but—in a country where a majority are illiterate—also reading out the content to the recipients.

Singh went to Mayo College in Ajmer and St Stephens in Delhi, and, after 25 years of academic break, attended the Maxwell School of Citizenship and Public Affairs at Syracuse University, New York. At Mayo, 'in true public school traditions, I played cricket, soccer, hockey, volleyball, tennis and TT, participated in athletics and gymnastics, went trekking, learnt leather, metal, clay and wood works, played the flute, did theatre and social service, debated and broke rules with impunity hard to imagine today,' his LinkedIn profile says.

Singh is invariably described by his colleagues as brilliant. His arguments even in casual conversations can be structured and logical—almost as if he is participating in a debate. He also tends to see, like a chess player, a few moves ahead of everyone else in the room. The corporate types from big consulting firms visiting his office often went with the impression that he was much ahead of them in the game—sometimes because he would make that clear in no uncertain terms. He can be brutally honest.

Singh not only immediately grasped the significance of a unique identity, he could see how it would align with his own wish list. He sought an appointment with Nilekani through an email. The response was prompt, and soon they were sitting across a table at a restaurant. 'I told two things to Nandan,' Singh said. 'First, that the Aadhaar number should also be an email address. A billion people automatically get their address. The second was that the Aadhaar number should also be the financial address of citizens, meaning you can send money to anyone using the number. Nandan said "no" to my first suggestion. He accepted the second.'

Singh joined Nilekani's team in 2010 to look after the payments system and financial inclusion. When he joined, he told Nilekani that he'd end up staying in UIDAI longer than Nilekani would. It turned out to be true. Not only that, he eventually headed the Post Office Payments Bank, which gave him the opportunity to implement some of the ideas he had for financial inclusion.

Singh often accompanied Nilekani when he had meetings that related to banking or finance. He not only had a deep theoretical, even philosophical, understanding of finance, he had practical knowledge from his post office days. (India Post also ran an extensive savings infrastructure).

At one point, UIDAI found that many were taking photocopies of Aadhaar letters as proof of identity and proof of address. Within UIDAI there were discussions on whether they should provide an option to do it digitally. This would mean getting into a whole new area of eKYC, whereas UIDAI's mandate was to provide an identification to all the residents and a way to authenticate it. Singh was clear. He said that UIDAI does not *own* the data of the residents. It's a *custodian* of residents' data. If they wish to share their data with another individual or entity, they should be able to do it. And it falls within UIDAI's mandate to provide that.

Years later, at an event in Bengaluru, Nilekani presented a framework called DEPA, short for Data Empowerment and Protection Architecture. The basic idea behind it was that your data belongs to you, and you should have the right to share your data for your own benefit. That's not how the world operates today. The quantum of data that a Google or a Facebook has about us, in terms of raw data, metadata, and derived data (crunched by artificial intelligence and machine learning) is orders of magnitude larger than the data that we have.

Diversity Can Drive Innovation—and Conflict

To build an organization that had a governmental scale at start-up speed, governmental stability and start-up agility, governmental process orientation and a start-up's sensitivity to markets, Nilekani knew that he had to embrace diversity at many levels.

The first level was a diversity from government services. UIDAI had people from various services, including IAS, IRS (Indian Revenue Service) postal service, financial service, railways, and defence accounts.

The second level was getting people from the private sector. To fill key positions such as technology, project management, and marketing, he hired laterally. For example, Srikanth Nadhamuni, who headed the technology centre in Bengaluru, and Shankar Maruwada, who headed marketing for the project, were hired as full-time employees.

After getting an engineering degree from the University of Mysore, and a masters from Louisiana State University, Nadhamuni had worked in some of the iconic tech companies in the US, including Sun Microsystems, Intel, and Healtheon (which featured in *The New New Thing*, Michael Lewis's account of the Silicon Valley). He came back to India in 2002 to do 'something meaningful for the country, instead of watching it from the US' and started eGovernments Foundation, along with Nilekani.

Maruwada's background was different. He got an engineering degree from IIT Kharagpur, and then an MBA from IIM Ahmedabad, spent four years at Procter and Gamble, before co-founding Marketics in 2003, which, as the name suggests, was into marketing analytics. Five years later, WNS bought the start-up for $65 million. Maruwada took a break to spend his time travelling and reading books. When the first TED Talk was organized in India, he was one of its sponsors, and he met Nilekani at an event dinner. The few minutes he spent with Nilekani made an impression; Nilekani said he had a role for him and would get back soon. That was to take care of marketing for UIDAI. Even at that time, Nilekani knew that it would be voluntary, and would need some marketing for it to be accepted by people at large.

There were other volunteers as well, who spent about a year or more working on specific activities. UIDAI also tied up with companies such as McKinsey, Intel, and Genpact for sabbatical programmes, letting their employees work in UIDAI for a specific period of time. The government already had internship programmes that UIDAI made use of.

Sahil Kini, an engineer from IIT Madras, took a sabbatical from McKinsey to join the programme. Sanjay Swamy, the CEO of a mobile fintech company called mCheck, joined as a volunteer. Some of them joined as volunteers so they could start working immediately, and then got absorbed into UIDAI as employees. Srikanth Nadhamuni, Pramod Varma, Sanjay Jain (who in his previous job at Google had helped create Google Mapmaker, and volunteered to work for UIDAI after listening to a talk by Raj Mashruwala, an early volunteer who helped design the system, and helped UIDAI access technology talent from the US and India), Vivek Raghavan (a biometric expert from Magma Design Automation)—all joined as volunteers, and some of them stayed long after Nilekani left.

Part of the reason why many queued up was the scale of the project's ambition and Nilekani's charisma. But there were bigger forces at play too.

In many ways, the world in 2009 and 2010 was very different from what it is today. The West was still licking its wounds from the financial crisis. The feeling that the private sector can go horribly wrong, that the checks and balances in place might not work, and that government would have to step in to save the economy from collapsing was pervasive across the world. People in general looked at government with kindness.

Meanwhile, the Indian IT sector had lost some sheen because the top companies, which were getting most of their revenues from the US and Europe, were feeling the heat. But analysts and industry observers trusted the top management to tide over the crisis. TCS was the largest software exporter, but the brand value of Infosys was much higher. (Cognizant was growing fast, but at that time, no one actually believed it would be a threat to Infosys anytime soon. People still believed in the books of Satyam, one of the top five software companies in the country; the company was tagged as 'India's Enron' after it emerged it had been falsifying its financial statements.[13])

In such circumstances, Nilekani's move to the government had triggered a surge of positive conversations. In those days, you couldn't attend a single conference without hearing executives discuss his new position. In some ways, they were reminiscent of the time when Nachiket Mor, one of the top executives of ICICI Bank, India's largest private sector bank, decided that he would opt out of the race to the corner office to head the bank's philanthropic foundation and focus on development.

A piece in the *Economic Times*[14] went like this:

> Inside ICICI's hothouse of competitive meritocracy, it was almost akin to MS Dhoni declaring a walkover before the final over of a T20 game with just 14 runs to get.
>
> Now consider this: there could be many, who may hate to admit it, but can't help feel a tinge of envy at Mr Mor's guts.
>
> The truth is that many corporate professionals today find themselves trapped inside careers that are no longer fulfilling or meaningful. Ask yourself this: in the last six months, how often have you woken up from sleep on a Monday morning and felt like rushing off to work? Again, the truth is that only a few can muster the courage to walk away from six-figure salaries, fancy mansions and fast cars. But thanks to the daring of those few, inside India Inc's gated community, it is increasingly becoming hard to wish away the Other India outside. Those who have made the crossover say development offers a much wider and more satisfying role than corporate careers can match. Success and money, beyond a point, make one wonder, what next?

Capitalism has a way of responding to the needs felt by its most important resources, and not long after, it was increasingly becoming the conventional wisdom that future leaders can't afford to focus on business alone.

Nilekani knew there was a desire among many in the private sector to contribute to government. 'The marginal benefit for a bureaucrat working for UIDAI as against working in any other department or project is insignificant. The life of a bureaucrat goes on,' one bureaucrat told us. 'But for a private sector person, to be associated with a project as ambitious as this, the marginal benefit is huge—even if the project failed to take off. It's like Pascal's wager [which in effect says when the losses are finite and the gains are substantial, go for it]. It made a lot of sense to trust in Nilekani and jump in. The real challenge with volunteers is not attracting them, but in creating space in the government to accommodate them.'

(In his book *Redesigning the Airplane While Flying*,[15] Arun Maira, former India chairman of Boston Consulting Group and former member, Planning Commission, narrates with gentle humour the difficulties he faced in bringing people from the private sector to work in the Planning Commission with its labyrinthine rules.)

Nilekani, though, came into the government with a much bigger halo than anyone else before. He held the rank of a cabinet minister. And, most of all, everyone in government knew that the kind of technological talent that was needed to build a project could not be found within the government. Though it was a necessary condition to execute the product, the government was not a sufficient condition.

Nilekani also knew that he had to rely on people from the private sector for another reason. The government machinery was slow. For example, while Nilekani first had a chat with Ganga as early as May 2009, it was not until August that she could actually join the team.

A Start-up on Steroids

In 2009, a couple of executives from a top global tech firm specializing in biometric algorithm landed in Bengaluru to give a presentation and demonstrate their capabilities to a bunch of people who said they were working on a project for the government of India. The address they were looking for was off the Outer Ring Road, a 60-km stretch around the city dotted with some of the biggest names in the tech industry. However, as they approached the address they saw it was an apartment—a sparsely furnished one, with a couple of desks, white boards, a refrigerator, and an oven, but no television set. There were four people (counting the one who briefly walked into the hall in shorts and a t-shirt, a laptop in his hand, before disappearing into one of the rooms. He hadn't taken his eyes off the laptop screen; it seemed as if he came out seeking an answer but found it somewhere on the way). The other three seemed very sharp; they asked pointed questions, and appeared to grasp the biometric experts' points extremely fast. Yet, the visitors' doubts grew bigger and bigger: was this bunch of people really with the government, or were they actually rivals doing some corporate espionage? Perhaps they took some comfort from the fact that if the three were really imposters, the house would not be so blatantly start-up-ish.

Yet, in the early days, UIDAI worked from this apartment that Nadhamuni, Raj Mashruwala (a Silicon Valley veteran who was associated with Tibco Software and Consilium), Pramod Varma (who had worked in Infosys, Yantra, and Sterling Commerce), and others would hang out to endlessly discuss technology for the project. Nadhamuni's home was close by and his help would prepare coffee. ('She would dutifully ask if we preferred coffee or tea, but bring us only coffee no matter what we said,' Mashruwala later joked.)

'We are all hard-core technologists,' Nadhamuni said. 'We are excited by big problems. That gets our adrenaline going.

The only thing we ask for are a few chairs and a few large white boards, and we are ready to go.'

As soon as Nilekani knew he was going to head the unique ID programme, he started picking the brains of a bunch of people. They included (besides Nadhamuni and Sriram Raghavan, T. Koshy, and Maruwada), Deepika Mogilishetty, a lawyer, and Devi Yashodaran, who helped Nilekani write *Imagining India*, and a few others. Before Nadhamuni found the apartment near his home, this eclectic bunch would meet in Nilekani's house. Eventually they moved to a larger office space on Outer Ring Road.

This team, too, not only drew from their experience, but also had long conversations with others. Maruwada, for example, went deep into the subject of identity itself, picking the brains of those who looked at it through various lenses. He spoke to Devdutt Pattanaik, an author and researcher who has written extensively on Indian mythology in the context of today's business and politics, and to Santosh Desai, CEO of Future Brands, who is also a columnist and the author of *Mother Pious Lady: Making Sense of Everyday India*.

Maruwada and his team travelled extensively to remote parts of India and realized that one of the strongest brands among the poor was the state emblem of India, featuring four lions from an ancient Indian pillar from the days of Ashoka, dating 268 BCE. In fact, they hit on the word '*aadhaar*' to denote the ID on one such excursion. Naman Pugalia, who had joined as a volunteer, was talking to a few men in a remote village called Mogiyathana in Rajasthan, one of whom remarked, '*Pehchaan hi toh jeevan ka aadhaar hai*' (Identity is the foundation of life). The word 'aadhaar' resonated with him and he called Maruwada as soon as he could.

Meanwhile, in the Outer Ring Road office, the technologists were busy building, testing, and refining the basic modules. In 2010, when we visited the office, it was collegial. One evening, at around 8 p.m., the entire office was buzzing with activity. There were ad hoc meetings in the corridor. Elsewhere, someone was excitedly debating on the nuances of iris scanners. As we walked past him, a young man suddenly dashed to a cubicle and dragged a man with a mop of hair and a bright smile to his workstation, where the two had an intense discussion. A few minutes later, the problem appeared to have been solved, and the troubleshooter went to solve another problem at another workstation.

This go-to guy turned out to be Pramod Varma, chief architect of Aadhaar. His colleagues sometimes refer to Varma as the 'royal architect', because his ancestors used to rule a part of Kerala. However, Varma grew up in a strictly middle-class setup: His mother taught at a school; his father, who grew up in royal opulence, ran a watch shop. Varma attended a Malayalam-medium school that charged next to nothing; a couple of his classmates didn't even have a home to go to (their address was 'behind the temple'). He eventually picked up maths and computer science in Hyderabad and Delhi, worked in Infosys, and by the time UIDAI launched, he was with a US-based company called Sterling and was considered one of the best architects of large-scale systems. (He wrote to Nilekani, offering to volunteer. Nilekani, not surprisingly, agreed.)

An executive from Mindtree, which was developing the enrolment software for UIDAI, told us that Varma was truly a whizz. Varma once walked into Mindtree's office when one of the team members was breaking his head over a problem late at night. When the engineer asked if he could help, Varma listened to the problem, told him that he was not familiar with that platform, but would see if he could find a solution. Next day, at around 8 a.m. he walked in with a dongle, handed it over, and said, 'Try this.' His solution worked like magic. When we recounted this story to Varma seven years later, he didn't make a big deal of it. 'The platforms might be different, but the principles are the same. So I guess it must have been easy,' he said.

UIDAI needed this eclectic mix of people for a number of reasons. One, building and deploying the system demanded inputs from a large number of stakeholders. Those who joined UIDAI might still not have all the answers, but at second degree of separation, there was an extremely diverse pool to tap into. Second, a lot of what UIDAI had to do was through persuasion by logic and appealing to self-interests, because there was no law mandating different state governments to co-operate. Even enrolment for citizens was voluntary—at that time, in both letter and spirit. Third, sooner or later, the identity infrastructure had to be used to provide services to people. Such diversity was needed to think of use cases from across sectors, and it also sent a signal that it had a large number of applications.

It had the desired impact. As government officials moved back and on to other departments, they went with a good idea of what Aadhaar could achieve. The well-known examples are that of Ram Sevak Sharma, who deployed an Aadhaar-based attendance system back in Jharkhand (which turned out to be one of the things that impressed Narendra Modi when he became the country's prime minister in 2014), and that of Ashok Pal Singh, who went on to head Post Office Payments Bank. There were other lesser known examples. Shrikant Karwa, who spent a few years as a volunteer at UIDAI, joined World

Bank's Identification for Development programme. Sanjay Swamy, a volunteer who was looking at fintech applications, started a venture capital fund.

But the diversity also meant clash of cultures. Initially, it was just funny. The government people—as they were referred to by those from the private sector—were bemused looking at volunteers walking around the office in shorts and T-shirts. Private sector people—as they were referred to—never understood why they had to get up like school kids when a higher ranked official came into the room, or why they should be addressed as 'sir' or 'madam', when they could call Nilekani 'Nandan'.

Some broke bureaucratic etiquette assuming that it had no logic. But there were reasons. Many private sector people were in the habit of marking a cc to Nilekani and Ram Sevak Sharma when they mailed the officials they were reporting to. Bureaucrats frowned upon that practice. As one of them explained, 'By virtue of getting cc-ed in a mail, they in some ways become a party to a transaction that is between two officials under him or her. If you are a party to a transaction, your authority to judge the transaction diminishes.'

Government officials were beginning to feel that while they were also putting in hard work, the other group was hogging all the attention. Government officials not only had to cross several hurdles before they got permission to speak to the outside world, their culture nudged them to be invisible. That would have been fine if the project itself were not constantly at the centre of media attention, but it was, and that led to some resentment.

Eventually, those from the private sector found themselves getting gently elbowed out of the system, thanks to bureaucratic capture. This change was felt even by bureaucrats. Ashok Pal Singh told us that for most of his career, he felt he was swimming against the tide. When he joined UIDAI, things changed. The whole organization was aligned to go forward, get things done. But within a couple of years, that old feeling returned. UIDAI was getting more and more bureaucratic.

Later, when we asked Nilekani about it, he said it was intended to be so. That unique organizational structure was needed to execute the project fast. After that it had to be owned by bureaucrats, and that's what happened, he said.

For some, it was clear even during the early days of UIDAI that the nature of the organization would keep changing. Speaking to us in 2010, Raj Mashruwala drew parallels to war. 'In the US, they say, "the Marines go in and then the Army takes over". Marines specialize in one kind of skill set, the army in another.'

Nilekani himself might have been influenced by his years in software services, where they make a distinction between application development and maintenance. UIDAI in the 'application development' mode needed one kind of organizational structure. By 2012–13, it had shifted into maintenance mode.

A Lean Organization

One of the things that Nilekani didn't want to do was to create a huge organization. If there is one thing that is worse than a bureaucracy, it is a technology bureaucracy, Nilekani said. 'If you are a bureaucrat, you get exposed to various services. And that gives you a broad view. In a technology bureaucracy, you are too focused on technology. The narrow exposure plus the law of bureaucratic gravity can be dangerous.'

Nilekani was looking to build an organization that was lean, strong on processes, and capable of creating systems that would make the best use of incentives that the broader ecosystem would respond to.

To do that within the government was difficult—but not impossible. In some ways, UIDAI was a start-up even within the government.

UIDAI was a part of the Planning Commission, and in the early years all they had was a single office room that the top three officers—Ram Sevak Sharma, B.B. Nanavati (who joined the UIDAI from the IRS), and K. Ganga—had to share. Eventually, Sharma got a corner in Nilekani's office, and the space constraints became a little less, but not by much. When an officer from Ganga's previous assignment came down to the UIDAI office to finish off some formalities with her, he was shocked by the size of the room. Her office in Shimla was grand. 'I am sure he went back thinking UIDAI was some punishment post,' Ganga said.

The hours were punishing. But they were literally creating a new kind of organization—one that would seamlessly work with the private sector—in effect, be a start-up on steroids. In start-ups, everyone does everything, and that was pretty much what the senior officers were doing. And in the process, they tried to make the system a little better. While government departments typically choose IAS officers based on a single metric (the overall assessment) that one can find on the last page of their file, Ganga went through every page of the submitted dossiers while recruiting, giving special attention to the officers' self-assessment to see if they would fit into the UIDAI culture.

The UIDAI culture at that time was to a large extent defined by multiple needs that the situation demanded. There was a need for speed, a need to maintain quality while accelerating, a need to be open to new ideas, and take risks—all of these within the boundaries dictated by the rules of bureaucracy. To a large extent, bureaucracy was designed and evolved to keep the machine running, and not to invent a new machine. But, here they had to do both. So, though she could not start with a blank slate, she could rearrange the elements to design a better system.

While hiring temp staff, government agencies typically told staffing companies how much they would pay for, say, an office staff. This often created a perverse incentive, because staffing agencies picked up unskilled people, paid them peanuts, and sent them to government offices. UIDAI inverted the process and set the staff's salary, say at Rs 9,000 a month, and asked the staffing companies to quote a single number—how much did they want from the government to cover their expenses plus their profits over and above that base figure? This system avoided a couple of major problems riddling the earlier system. Since the salary was fixed, the staff did not get short-changed. Since the salary was good, the employee turned out to be good. Most of all, the simplicity of quoting a single number made the entire process effective and transparent.

In Bengaluru, similarly, the tech team designed a system that would incentivize tech partners. At the core of Aadhaar's biometric authentication is a de-duplication algorithm, which figures out whether each biometric image is unique. It is huge, in terms of both scale and complexity. To give the billionth Aadhaar number, it has to check a person's biometric data for uniqueness against 11.99 billion biometric images (fingerprints and irises). No system is so perfect that it will generate zero error rates. UIDAI enrolled three different vendors to do the task, each using their proprietary algorithm. Those who did better were dynamically allocated more work and thus more money, creating competitive pressure.

Bureaucracy has long believed in creating elaborate structures to execute projects, and even more elaborate structures to provide checks and balances. This approach has a great intuitive appeal, even if on the ground it had over and over again led to cost and time overruns, and a garbage can full of failed, stalled, mothballed projects.

What the bureaucrats in Delhi and the technologists in Bengaluru were aiming for was to create an incentive structure that would ensure speed, scale, and quality. The enrolment agencies were paid based on the number of Aadhaar numbers generated, and not on the number of enrolments. Which meant that they had to ensure that they paid attention when registering people.

A similar inbuilt incentive structure also helped the project avoid false information, a problem that riddled many other ID projects. Because it was a lifelong ID, and the only one a person would ever get (within the constraints of biometric technology), there was an incentive to give correct information. 'If you enrol yourself as Amitabh Bachchan you will be stuck with that name all your life. It's as simple as that,' Nilekani once said. A man from Delhi tried to get two Aadhaar IDs, enrolled once with his real name, and a second time with a false name. It so happened that his second enrolment got processed first, and the first got rejected. He is now stuck with an identification with a fictitious name.

This approach allowed UIDAI to remain lean, but that came with its own problems. If there is one lesson that economics teaches again and again, it's that incentive systems are hard to design. Incentives can turn perverse. *Freakonomics*,[16] a popular economics book that explains how incentives work using a number of quirky examples, tells the story of a child care centre that was facing a common problem. Parents came too late to pick up their children. So, the centre decided to impose a huge fine for late pickup. The move had an impact, only in the opposite direction. The late pickups went up. Rather than making the parents come on time, the huge fine turned out to be an excuse—a legitimate fee—for coming late. Parents who were trying to come on time, out of a feeling of guilt for keeping the staff waiting, now felt they had earned the right to come late.

Even a well-thought-out incentive structure might not work really well. To ensure that enrolment agencies do their best to capture biometrics of residents, UIDAI paid them based on the number of Aadhaar numbers generated rather than the number of enrolments. To beat that, a number of enrolment agents started using the biometric exceptions. A set of agencies in Andhra Pradesh enrolled thousands of people using this route. Incentives alone didn't work. By 2017, UIDAI had suspended about 50,000 enrolment agencies.[17]

Anand Venkatanarayanan, a cybersecurity professional who has looked at UIDAI through the lens of security and processes, says that in its attempts to stay lean, its ability to respond to security issues got severely undermined. UIDAI does not have a chief information security officer, and it has launched a number of products with security vulnerabilities. No system is 100% secure, but the way UIDAI has been responding to security flaws in the system is not reassuring at all, he said. 'UIDAI is essentially a shell company that takes money from the government and outsources it to third parties. It is a structural problem,' he says.

Criticisms such as these, while being correct in recognizing that there is a problem, tend to mistake issues with control for issues with structure. The outsourcing wave that swept over global manufacturing and IT services shows that it is possible to have control over key processes without having to own the resources and employees.

You can't create a perfect system, says Teamlease's Manish Sabharwal. 'What you need to look at is whether the weaknesses are prejudiced, pervasive, and permanent. Does a system have an inherent bias against a certain set of individuals? Whether the weaknesses are pervasive across the system, or localized. Whether the weaknesses are permanent, or if they can be set right. In my view, they are not,' he says.

The question then is whether the organization can evolve and keep evolving. There is evidence to show that it's happening. You have to look at their new updates to get a sense of it. They'd introduced the biometric lock, virtual Aadhaar, face–fingerprint fusion authentication in the recent weeks and months, Sabharwal pointed out.

Part of the reason for this evolution is that India is a democratic country and politicians cannot hope to win elections by turning a blind eye to the problems in the system. It has a strong and vocal civil society, constantly putting pressure. Also, UIDAI hadn't entirely cut off its connections with the people who built the system. Pramod Varma, for example, continues to be an advisor to UIDAI, and has been pushing for some of these upgrades.

First among Equals

On 29 September 2010, just 16 months after Nilekani was appointed chairman of UIDAI, 30-year-old Ranjana Sonawane became the first person to get an Aadhaar number. Sonawane stood on stage, flanked by India's Prime Minister Manmohan Singh and UPA chairperson Sonia Gandhi.

The government had picked up a tribal village to distribute the first cards, to highlight that it was an inclusion project. Nine others, all of them poor, most of them day labourers, also received their cards with their unique 12-digit number from Gandhi, the president of the ruling party. Gandhi gave a short speech. 'The aim of the UID [universal identity] scheme is to bring transparency in the system,' she said. 'Even if an individual migrates or shifts temporarily to some other place, she or he will be entitled to benefits of all government schemes. The UID holder will not have to run from pillar to post for establishing his or her identity and other details. I hope this project will change the life of every individual.'

Nilekani was on stage too. He was dressed not in a business suit as he used to during his Infosys days, but in a white kurta-pajama and a Nehru jacket. In one photograph taken on that day, Sonawane is flashing her smile, Sonia Gandhi has a triumphant laugh on her face, Manmohan Singh has a satisfied grin—and Nilekani looks humble, his hands locked in front of him, head slightly bowed looking at the Aadhaar card in Sonawane's hands.

Was he thinking, 'One step taken. A billion more to go'?

Notes

1. ID4D, 'Global Dataset', World Bank, 2015, https://datacatalog.worldbank.org/dataset/identification-development-global-dataset, viewed on 27 June 2018.

2. Sustainable Development Knowledge Platform, 'Sustainable Development Goal 16', United Nations, https://sustainabledevelopment.un.org/sdg16, viewed on 27 June 2018.
3. PTI, 'Multi-Purpose National ID Cards Distributed in Capital', *Hindustan Times*, 27 May 2007, https://www.hindustantimes.com/delhi-news/multi-purpose-national-id-cards-distributed-in-capital/story-UPypZae688fACN2EIaTxeM.html, viewed on 27 June 2018.
4. Raghuram Rajan, 'A Hundred Small Steps', http://planningcommission.nic.in/reports/genrep/rep_fr/cfsr_all.pdf, viewed on 27 June 2018.
5. John Kingdon, *Agendas, Alternatives, and Public Policies* (Boston: Little Brown, 1984).
6. Steve Blank, 'What's a Startup? First Principles', 25 January 2010, https://steveblank.com/2010/01/25/whats-a-startup-first-principles/, viewed on 15 July 2018.
7. Russell Cropanzano and Maureen L. Ambrose (eds), *The Oxford Handbook of Justice in the Workplace* (New York: Oxford University Press, 2015).
8. YouTube, 'The Power of One', https://www.youtube.com/watch?v=_QzjqOl2N9c, viewed on 27 June.
9. Walter Isaacson, *The Innovators: How a Group of Hackers, Geniuses, and Geeks Created the Digital Revolution* (New York: Simon and Shuster, 2014).
10. Joshua Wolf Shenk, *Powers of Two: Finding the Essence of Innovation in Creative Pairs* (London: Houghton Mifflin Harcourt, 2014).
11. Shenk, *Powers of Two.*
12. Anita Joshua, 'Tharoor's "Cattle Class" Tweet Annoys Congress', *The Hindu*, 16 September 2009, http://www.thehindu.com/news/national/Tharoorrsquos-cattle-class-tweet-annoys-Congress/article16881830.ece, viewed on 27 June 2018.

13. *The Economist*, 'India's Enron', 8 January 2009, https://www.economist.com/node/12898777, viewed on 27 June 2018.
14. Indrajit Gupta and George Smith Alexander, 'The Other India Inc', *Economic Times*, 3 November 2007, https://economictimes.indiatimes.com/news/company/corporate-trends/the-other-india-inc/articleshow/2513916.cms, viewed on 27 June 2018.
15. Arun Maira, *Redesigning the Aeroplane while Flying: Reforming Institutions* (New Delhi: Rupa, 2014).
16. Steven D. Levitt and Stephen J. Dubner, *Freakonomics: A Rogue Economist Explores the Hidden Side of Everything* (New York: William Morrow, 2009).
17. PTI, '50,000 Aadhaar Enrolment Operators Suspended till Date: MoS IT', *Hindustan Times*, 29 December 2017, https://www.hindustantimes.com/india-news/50-000-aadhaar-enrolment-operators-suspended-till-date-mos-it/story-LzLlKs0Yh9PxfeupFfiY9K.html, viewed on 27 June 2018.

2

The Art of War

'TELL ME A STORY, MOMMY,' THE CHILD PLEADS.

'Okay, what story do you want?'

'I want a Vikram–Baital story,' she says. She likes this story-puzzle series from Indian mythology. 'But a new one.'

And so the mother begins.

'King Vikramaditya, who had promised a sorcerer that he would capture a vampire, climbed the tree the Baital dwelt on.

'As the king carried the Baital on his shoulders back to his kingdom, the Baital said, "O king, I will tell you a story, and then ask you a question. If you know the answer and still don't give it, your head will burst into a hundred pieces. So listen carefully.

"Once upon a time there was a big city. None of the houses there had numbers. The postman had a tough time delivering letters. He had to knock on several doors, ask several people to find the right address. Sometimes, he couldn't deliver the mail at all. Sometimes, people lied and took letters meant for others.

"One day, a wise girl came up with an idea. Why not put a number on every door? Then the sender need only put the right house number on the letter. The city administration thought it was a good idea. Assigning a house number and keeping a record of it would solve the problem. It could even help others find the right address. It might even solve property disputes.

"But no sooner had the project started, different groups of people started finding fault with the idea and the way it was being done.

"The milk suppliers union said they had a superior solution, for they too had faced a similar problem: often two different people went to deliver milk and collect the bill, and the latter had trouble identifying the houses. Their solution not only identified the house but also the household's milk requirement. So, writing a number on the door was duplication of effort and the people should wait for the milk suppliers' project to finish.

"The city's financial controller said this would put a strain on the city finances. 'The cost of paint!' he exclaimed, running through a row of numbers in his ledger while shaking his head. 'Why don't they use chalk?'

"The businessmen who made nameplates had their own concerns. Assigning just a number is a bad idea; it should be on a beautifully designed board. It so happened that they had the exact kind of board.

"A group of well-meaning social workers—who loved the idea of postmen delivering letters to numbered houses, but were uncomfortable with this new idea—rushed to the debate. What if, they asked, some houses get left out during the process? What if it's a hut without a proper door and the numbers are not visible? 'It's a bad idea. Scrap the project,' they campaigned.

"Another group rushed in to say that the city is assigning numbers only so that it can keep an eye on the people. It can do surveillance even now, of course, the group leader conceded.

But the house numbers make it much easier. 'The government is bad already; why should we make it easy for them?' he asked.

"One small group assembled in a secret basement. 'Until now we had free access to some packages because no one knew where they were supposed to go. What do we do now?' a young blood asked. 'Maybe we can ask politicians, lobbyists, PR to speak up for us. Or else, lie low and pray that those who are against this project for their own reasons succeed,' said an old man.

"Another group of people who had nothing to hide met in a coffee shop to discuss how things could go bad. Someone said, 'We can't trust the city to do anything well. Imagine if two houses get the same number!' Someone else said, 'Imagine if someone intercepts a letter. All he has to do is change the house number on the letter to his own house and the letter will be delivered to him. This system is riddled with holes!' An older man got up and said, 'Privacy is important to me. I wouldn't let my house number be known to anyone.' They all decided to campaign against the project.

"Journalists went ahead to capture the brewing conflict. A politician quickly calculated that he could score some points against those in control of the city administration and started collecting people's grievances and fears about the project—and started putting pressure on the administration.

"The city administrators took note of all the opposition. They called for an emergency meeting to discuss the project. It had its benefits, but seemed to carry unspeakable risks too. They needed to decide what to do about it."

'There was a long silence.'

'And then, Baital asked Vikram, "Tell me, O king, what should they do? If you know the answer and still don't tell me, your head will burst into a hundred pieces."'

The mother looks at her child, and says, 'Now, tell me, what should Vikram tell Baital?'

The child is intelligent. This was like no other Vikram–Baital story she had heard. Spotting the newspaper lying in her mother's lap, she asks, 'Mommy, what is this story really about? Is this about Aadhaar?'

'No, my child. Aadhaar is far more complicated. I will tell you that story another time. Now, how would you answer Baital?'

The child thinks for a while. 'I don't know the answer, Mommy. But I know the moral of the story: There is no journey that you can undertake without facing obstacles.'

The Person and the Situation

A system in a state of equilibrium tends to remain so. As human beings, we like status quo. We normalize what's common, even if it is not strictly fair to all. We don't get sufficiently shocked by poverty, hunger, violence, inequality (or corruption or traffic violations). It's just like how our sense of hearing mutes familiar ambient noise and amplifies strange noises; or how our sense of smell makes us anosmic to familiar smells and very sensitive to new ones. But we are outraged when we hear about the slightest of deviations. When a policy or an action tries to push a system in a new direction, when it disturbs the status quo—even if it's for the greater good, even if it's to reduce hunger and poverty—multiple forces rise to push the system back to its equilibrium.

UIDAI'S objective was simple: give a unique identification number to all Indian residents. Given the size of the country, it was a massive project, with a potential budget of billions of dollars. Gradually, it was seen as a game changer for many, including the government, the social sector, and businesses. It promised huge benefits. But, there were also huge risks. Opposition started intensifying.

The most public opposition came from the Ministry of Home Affairs, which was building its own registry. It had some elements of the sibling rivalry that played out between billionaire brothers Mukesh and Anil Ambani, and caught much media attention. But UIDAI also faced challenges from the Planning Commission—a case of maternal envy; from businesses, which had issues with some of the design decisions UIDAI took; from the political establishment, which found an opportunity for one-upmanship; from NGOs for ideological reasons; from interest groups, because they stood to lose out from Aadhaar-based reforms; and from a group of individuals fighting it in the courts on various legal grounds.

Nilekani knew it could distract from getting things done. He wanted, above all, momentum. He had made a pact with Ram Sevak Sharma, the CEO of UIDAI. Sharma would focus on getting things done within the organization, and Nilekani would manage the ecosystem—in effect, shield the team from external pressure. That job would turn out to be far tougher than he had ever imagined.

Manish Sabharwal, the co-founder of staffing services firm Teamlease, maintains that there are two views of history. 'One is the sociologist view, which says things happen because of circumstances, because of larger political, social, economic, and technological forces. And then there is the literature view of history, which says that things happen because of a few good men. I believe in the literature view of history. I think Nandan is a great example of that. A lesser person would have succumbed to the challenges from multiple fronts. But Nandan didn't. He held his ground, and took his opponents head on.'

Nilekani himself, one may guess, believed in a bit of both. When big forces converge, they create a policy window, like John Kingdon explains in the political science classic *Agendas, Alternatives, and Public Policies.*[1] And then it's for the policy entrepreneurs to make the best use of the opportunity. Or as Sun Tsu said in *The Art of War*,[2]

> The general who wins a battle makes many calculations in his temple before the battle is fought. The general who loses a battle makes but few calculations beforehand. Thus do many calculations lead to victory, and few calculations to defeat; how much more no calculation at all!

A war never ends easily. To this day, those who are closely embedded in the system continue to be in battle mode. They bring in new recruits. They come up with new war cries, even as they fight the old battles. UIDAI has survived, grown bigger, and continues to engage in tough battles.

As for Nilekani, even though he no longer plays any official role in UIDAI—or for that matter in the government—he is often pulled into the battle. He doesn't say no. He is emotionally attached to Aadhaar. He sees it as his legacy.

In an interview in December 2017, Nilekani told us that he toughened up during the time he was in government and politics. 'I dealt with such complex situations, and such "bad actors" … In some sense a corporate situation is a piece of cake; intellectually it's not complicated. In government it was people with their daggers drawn. You won't meet them in a dark alley. When I was younger I was not good at handling confrontation.

I tried to avoid confrontation. I would acquiesce to something because I didn't want confrontation. But the big thing about the last ten years was that I have become comfortable with that. Not that I seek confrontation, but it doesn't bother me. Actually it makes me stronger to take on adversaries.'

The Turf Wars

The bureaucrats at the Home ministry had every reason to look at UIDAI with suspicion. UIDAI, an upstart with no statutory status, was seen as getting onto their turf.

First, giving a national identity card was their business much before UIDAI was formed. India started to seriously look at an identification programme back in 2002, after the Kargil War between India and Pakistan from May to July 1999. Atal Bihari Vajpayee was the prime minister then, and a group of ministers—that included then Finance minister Yashwant Sinha, External Affairs minister Jaswanth Singh, and Defence minister George Fernandes, under the leadership of then Home minister Lal Krishna Advani—recommended that Indian citizens get a Multipurpose National Identity Card.[3] The job was entrusted to the Home ministry. UIDAI on the other hand was formed only in 2008.

The Home ministry's programme was backed by law. In 2000, the Vajpayee government had amended the Indian Citizenship Act, 1955, and later passed another act in 2003—the Citizenship (Registration of Citizens and Issue of National Identity Cards) Rules.[4] These two made it mandatory for citizens—or usual residents—to get themselves registered in the National Population Register (NPR).

UIDAI, on the other hand, made a lot of hue and cry about it being voluntary, but for a long time was not backed by a legislation that was passed by parliament.

The Home ministry's programme had a broader scope. It would take care of security—in fact, it was born out of concerns for national security. The database could also be used for better targeting of government benefits and reduce identity fraud. Thus, the bureaucrats in the Home ministry argued, it was more efficient use of government funds.

The ministry was also collecting more data. It was collecting 15 data points about 'usual residents'. And its data collection was more robust. It actually did door-to-door surveys, and, the Home ministry argued, did not depend just on questionable systems such as introducers. Aadhaar allowed those with no documents to enrol themselves through this method. NPR did not use third-party enrollers to collect data, and looked at them with suspicion.

However, the Home ministry did not altogether dismiss UIDAI's efforts. The bureaucrats there recognized that biometrics had some value, and they were fine with the numbers that the system would generate after biometric deduplication. In short, they saw Nilekani as a techie who came to solve a back-office problem. As long as he acknowledged his role and served the needs of the Home ministry, they were fine.

But Nilekani had a different world view altogether.

He was betting not on the 'first mover' advantage, but on the 'fast mover' advantage. The Home ministry might have been working on an identity programme for a long time, but it was doing it slowly. A pilot for the multipurpose national identity card that began in 2003 took four years to finally issue the first set of cards to residents in a locality in Delhi.

As he saw it, an identity programme that was trying to recognize citizenship, address security problems, and also help the government provide better delivery of subsidies was simply doing too much. A good identity platform should do one job, and do it well. Provide an identity to a person—let her/him establish who she/he claims to be.

Shrikant Karwa, who worked as a volunteer with UIDAI and later joined the World Bank's Identification for Development initiative, said that one of the hardest things to communicate was the difference between foundational ID and functional ID. 'We are familiar with functional IDs, we have lived with them all our lives. We use [a driver's] licence as an ID proof, but it's a functional ID. It says that you are eligible to drive. A passport says you are eligible to travel outside the country. But a foundational ID doesn't establish anything. It's foundational.' This simplicity, this limited functionality, had its strengths. It can be used anywhere. (Many critics found it hard to grasp this feature, and accused Nilekani of selling Aadhaar as everything, to everyone.[5] Aadhaar, as we will see, is a technology component, a Lego block that can be used to build solutions across sectors—government, social, and business.)

The Home ministry was priding itself on collecting more data. Nilekani's team was fighting a battle to collect as little data as possible. Pramod Varma, the architect of UIDAI, said that simplicity—having to collect less data—was key to scaling up. Every additional data point meant a billion more data entries—additional time, additional effort, higher possibility of introducing an error. But more than that, the more data you want to collect, the less inclusive you will make the system. 'If you want to make father's name and mother's name a mandatory field, what happens to orphans who might not know who their parents are?' Varma explained. 'In government we tend to do certain things because we have always been doing it that way. But here we were asking what else to cut, what else to remove.'

Among those who were building Aadhaar, there was a Steve Jobs-like obsession to cut down, reduce, simplify. (Steve Jobs was still around when they were designing Aadhaar, and iPhone was hardly three years old). The predominant driver, however, was not aesthetics; it was values-driven, strategic pragmatism. 'We had our guiding principles—inclusion, inclusion, and inclusion,' Varma said.

Years later, Varma met a group from Afghanistan that had come down to India to study Aadhaar. They wanted to know what India did right with its unique identity programme, while theirs—which was trying to solve the problem at a much smaller scale—was struggling even to inch forward. Varma wanted to know what they did wrong, and got his answer when one of the delegates said they were collecting as many as 50 data points. Like a cart with too much load, it sank to the ground.

Similarly, the Home ministry was doing a door-to-door survey—while UIDAI was hoping to give Aadhaar to people who did not have a house to stay in.

The Home ministry's process was robust. It was creating a verified database. In fact, at the village level, the data collected for the NPR was displayed in public for verification. Nilekani and his team saw that as a complicated process that also raised questions about privacy.

UIDAI and its designers depended on the inbuilt incentive that a biometric deduplication injected into the system. Your biometrics are unique, and your Aadhaar number is going to be linked to your biometrics, so you will get only one

Aadhaar number. Therefore, it's in your interest to provide the correct details.

For Nilekani, momentum was the key. He wanted UIDAI to give Aadhaar numbers to 600 million people—or half of India's population—before his term ended. It had to become too big to fail, and it had to become that fast. Loss of momentum meant death for the project.

But the Home ministry was not one to give its turf away. No doubt the bureaucrats in the ministry knew their place in Delhi; in a bureaucracy where hierarchy matters, the Home ministry was among those at the top, measured by the number of joint secretaries that worked there and by the veto power it had (top appointments had to be vetted by the ministry). But it would also be unfair to see their displeasure as just power posturing. After all, the Home ministry's mandate on citizenship and national security is clear. And it was the bureaucrats' job to fulfil that mandate.

They couldn't have found anyone better than P. Chidambaram to fight their battle. This was in 2010. The UPA, often called UPA 2 because many expected its social-sector reforms to continue, had been re-elected in 2009 to form a coalition government with Manmohan Singh as the prime minister again. Chidambaram was the Home minister (since 2008, after terrorist attacks in Mumbai) and one of the most powerful members of the Congress party, the largest party in the alliance, not least because of his intelligence and legal acumen. He is sometimes described by those who know him as a 'super bureaucrat' because of his obsession with procedures and his ability to wear his opponents down with obscure rules and sub-rules, all of which he could access from his formidable memory.

The Home ministry had questioned the need for UIDAI right at the time of its formation—even before Nilekani joined the government—and was somewhat pacified by the assurance that UIDAI would not be looking at citizenship (which would remain in the purview of the Home ministry). Later, after Nilekani joined the government and gave his first presentation in August 2010, Chidambaram was not present, but made it a point to write to Nilekani about the superior process the Home ministry had put in place.

But it was not until UIDAI had almost finished enrolling 200 million people that the Home ministry came down heavily with its full force and stood in the way of allowing it to do the remaining 400 million enrolments that Nilekani was hoping for. Whenever Nilekani could seize an opportunity, he would tell Sonia Gandhi and Rahul Gandhi, the then president and vice president of the Congress party, respectively, about the impasse and request them to try and solve the problem. But their view at that time was that it should be resolved within the government.

'I think Nandan did not want to be in a situation where he was seen as close to one politician and against another, but the opposition from the Home ministry made him move closer to Pranab Mukherjee,' a senior bureaucrat said. Mukherjee was the Finance minister then and the most senior cabinet minister in the government; and by some accounts, he was practically in charge of most cabinet meetings. In the early days of Aadhaar, he had invited Nilekani to dinner and signalled to him that Nilekani should just do his job, and he would watch his back. He did.

But a few more things had to get aligned before the issue was resolved. The Prime Minister's Office (PMO) got a new principal secretary in Pulok Chatterjee, who understood the importance of digital identity as few others in the government could, thanks to his previous work at the World Bank (which had deep interest in innovation in development) as executive director. Prior to that, he had served in Rajiv Gandhi's PMO, in Rajiv Gandhi Foundation, and was Sonia Gandhi's officer on special duty when she was the opposition leader. He began pushing the needle.

In early January 2012, Prime Minister Singh called for a meeting between the warring factions (not unlike Kokilaben Ambani, widow of Reliance Industries' founder Dhirubhai Ambani, who intervened in the fight between her sons Mukesh and Anil over the control of the empire her husband built[6]) and persuaded them to agree to a compromise. And that was to divide the population into two segments, and give one half to UIDAI and the other to NPR. Both would share the data. NPR mostly got the border states where security and infiltration were a concern, and UIDAI got the others.

In effect, the Home ministry's fight to protect its turf was seen in the media as a fight between the two most senior ministers in Manmohan Singh's cabinet.

Speaking to us in 2017, Montek Singh Ahluwalia, former deputy chairman of the Planning Commission, made light of the fight between UIDAI and the Home ministry, saying it could have been resolved between Nilekani and Chidambaram over a cup of coffee. However, it was not just two individuals fighting with each other.

Events that unfolded later would show that the differences were indeed from structural reasons.

Mukherjee became the president of the country on 25 July 2012. Chidambaram left the Home ministry to become the Finance minister. As Finance minister, he saw huge value in Aadhaar, and towards the end of UPA rule, would become a strong proponent himself, batting for it in the cabinet when it faced another threat.

However, the fight between the Home ministry and UIDAI continued. When the BJP-led NDA was voted back into power in May 2014 after 10 years of UPA rule, with Narendra Modi as prime minister, almost everyone assumed that UIDAI would merge into NPR, thanks to the loud campaign against Aadhaar during the elections. In one of the first speeches that Modi gave, he said that many things might have been said during the campaign, but it was now time to govern. This was seen as an indication that he would not kill Aadhaar. And in one of his first meetings, Modi light-heartedly reminded Home minister Rajnath Singh that Aadhaar came directly under the prime minister himself, as he was the chairman of the Planning Commission—under which UIDAI was housed. The bureaucrats in the Home ministry realized that UIDAI would remain independent after all.

None of this, however, has stopped the system from carrying out its campaign or at least from hoping that Aadhaar would become a part of the Home ministry. Its officials continue to complain about the risk that Aadhaar poses for national security in their conversations with activists and journalists. That got amplified during the case now in front of India's Supreme Court to decide on the constitutional validity of Aadhaar.

But we are getting ahead of the story.

The Empire Strikes Back

If the Home ministry saw UIDAI as an upstart encroaching on its turf, the erstwhile Planning Commission (it was replaced in 2015 by the Niti Aayog) saw it as a unit—a subset of itself—going out of control.

The bureaucrats in the Planning Commission had all the reasons to keep UIDAI under its tight control.

UIDAI had an interesting history that stretched farther than before even Nilekani became a part of it. In 2004, the UPA came to power on the promise that it would ensure that the poor and the needy got their entitlements from the government, in contrast to its rival and predecessor NDA's India Shining campaign, which it said, shone only on the rich. As a follow-up, it announced a project to provide a unique ID for Below Poverty Line (BPL) families. The Planning Commission worked on a detailed plan, through a committee under economist Arvind Virmani, who later served as the chief economic advisor to the Indian government. The Commission also joined hands with IT services firm Wipro to prepare a strategic vision for the project.

The government formed an empowered group of ministers to put a system in place, and early in 2008, it decided that it would form a Unique Identification Authority of India. By November that year, UIDAI was set up as an attached office of the Planning Commission. Subhash Pani, secretary, Planning Commission, had envisioned a grand structure in line with the Election Commission—with a slew of joint secretaries and a UID Authority for every state. It would make use of the existing databases—primarily the electoral database—and work with the Home ministry to provide unique identities.

When the government announced that UIDAI would be headed by Nilekani, many within the bureaucracy didn't take him seriously. Nilekani himself was not a hundred per cent confident then. When Pramod Varma met Nilekani in 2009 to discuss leaving his job to join UIDAI, Nilekani warned him that all of this could come to nothing. Varma said he'd still join because he was 'one of the most optimistic people in the world' and because, if worse came to worse, he could always go back and find a job.

Varma and others who had joined Nilekani in the early days were confident enough in their own abilities to take the risk. But they were also self-aware. They knew many within the bureaucracy and political establishment looked at them as jokers who would make some splash in the media initially and soon become irrelevant.

Economist Paul Ormerod wrote in *Why Most Things Fail*:[7]

> Failure is all around us. Failure is pervasive. Failure is everywhere, across time, across place, and across different aspects of life. Ninety-nine point nine nine per cent of all biological species which have ever existed are now extinct. Failure in this context is measured over hundreds of millions of years …
>
> The Iron Law of Failure appears to extend from the world of biology into human activities, into social and economic organizations. The precise mathematical relationship which describes the link between the frequency and size of the extinction of companies, for example, is virtually identical to that which describes the extinction of biological species in the fossil record. Only the timescales differ.

People who eventually find success often use a quote, sometimes attributed to Mahatma Gandhi: 'First they ignore you, then they laugh at you, then they fight you, then you win.' But, in fact, it's a heuristic theory that has some basis in the way things work. Most things fail, and to ignore people who are trying to do something new, in a new setting, is one of the most efficient uses of our time.

Nilekani made the best use of that situation. Because the system did not feel threatened by his presence, he got room to hire his own core team (a luxury many in the government don't get). The team—R.S. Sharma, K. Ganga, and M.S. Srikar—were acutely aware of the bottlenecks in bureaucracy. They got Nilekani to get Prime Minister Singh, who was also the chairman of the Planning Commission, and its deputy chairman Montek Singh Ahluwalia to delegate some of their powers to him. This meant that the approvals did not have to go through the Planning Commission bureaucracy.

Nilekani had zero interest in turning UIDAI into the large bureaucratic organization that Pani had envisaged. He wanted it to be small and nimble. He was not impressed with the plans that UIDAI had prepared for giving identities. He wanted a blank slate—and got some of the best minds in the world to fill it up through public consultations, and volunteer and sabbatical programmes, along with the best minds from within the government. He spent the first few months getting inputs from some of the smartest people in technology and public policy as well—including T. Koshy, Srinath Raghavan, Nachiket Mor, Anil Jain, Esther Duflo, and Abhijit Banerjee.

He also had his technology team based out of Bengaluru, instead of New Delhi, not only because it was easier to hire technology talent in Bengaluru, but also because the distance from New Delhi would mean fewer distractions.

All these helped. By 2010, the first Aadhaar numbers were given. UIDAI had the clearance to enrol 100 million people, and it was going full speed.[8] Nilekani knew that he could not afford to lose momentum. He went to the Finance ministry and got approval to enrol the next 100 million, essentially side-stepping the secretary of the Planning Commission.

Bureaucrats in the Planning Commission didn't take these incursions kindly. (Bureaucrats are endowed with negative power—any bureaucrat can say 'no' or just sit on a file without giving a response—and bring activities to a halt. But bureaucrats are mostly tactical, and rarely strategic.) Nilekani kept his feedback loops open; he spoke to everyone from a secretary to a peon and knew how to wait. He also knew that some resistance came from individual bureaucrats and not the department as such. Thus, a bureaucrat who is fighting vigorously (and bureaucrats can fight vigorously by merely sitting on a file, promising to do something, and following it up with inaction) will retire in six months. So, he would simply wait for the person to retire and deal with the next bureaucrat who occupied the chair.

Bureaucrats also help each other out. One Friday evening, Nilekani and his team heard from the Comptroller and Auditor General (CAG) of India's office that it would investigate Aadhaar's books the following Monday. Nilekani was extremely careful about legal and financial issues because he knew that even a whiff of impropriety might undermine credibility and slow down momentum. He spent from his own pocket for his weekly trips to the technology centre in Bengaluru. But he also knew that words like 'investigation' have a negative connotation.

Many in the bureaucracy could quickly connect the dots. Sudha Pillai, secretary at the Planning Commission, who didn't like the idea of UIDAI side-stepping her for its budget (and whose husband was the home secretary, with whom UIDAI had an ongoing fight) and Vinod Rai, the CAG, were from the same batch. Bureaucratic caste rules suggested that same batch officers often stand up for each other.

This didn't go unnoticed by Shekhar Gupta, then editor-in-chief of the *Indian Express*, who wrote about the power play by bureaucrats. After describing the incident, he noted:[9]

> We have no evidence at all of the two working in concert in any way, and so let us simply presume that each one has been acting on his/her own, driven by bona fide doubts about the project. But what is evident is that each of them is at least individually challenging a programme so precious not only

> to this government, but also to the UPA's top leadership, the Gandhis. Let's not take a position on whether this is a good or a bad thing. Let us simply say that such things are nearly unprecedented in New Delhi.

Gupta ended his column with a warning. Referring to Robert Blackwill, a former ambassador to India and a professor at Harvard University, Gupta noted:[10]

> Bureaucrats, he [Blackwill] said, were like doctors and nurses in the emergency ward of a hospital. When a patient was wheeled in, their job was to follow Standard Operating Procedure (SOP) and wait for the specialists (the political leaders) to arrive next morning. The politicians would then decide what to do with the patient, and the bureaucrats would, in turn, implement those instructions. But you can imagine what would happen if people trained to follow SOPs took over the job of the specialists in a country as challenging to govern as India. The result would be a government in deep freeze, incapable of taking decisions, running on mere SOPs. In brief, a government in the emergency ward, which is as good a way as any to describe UPA 2 today.

Gupta's point about overreach of the bureaucrats sent a strong message—if not to the broader bureaucratic community, then at least to the people who mattered. Gupta had framed the fight as one between politicians and bureaucrats. Nilekani's legitimacy came not from his having passed the civil services exam and serving in the government for decades, but from the fact that he was invited by the prime minister, and because the prime minister and the party leadership believed the project would help them serve their constituents better.

But there was no denying that the Planning Commission held significant power, and nowhere was this power as visible as it was in the meetings of the National Development Council, the apex decision-making body on development issues, that were held in Delhi. The Council includes the prime minister, cabinet ministers, chief ministers of all states, and the members of the Planning Commission (now Niti Aayog).

In his book *An Upstart in Government*,[11] Arun Maira, former India chairman of Boston Consulting Group (2000–08) and former member, Planning Commission (2009–14), described it thus:

> The conduct of the National Development Council's meetings angered many chief ministers. They were made to sit in the well of the stately Vigyan Bhavan while the Chairman of the Planning Commission (the prime minister) sat on the stage above them, along with the deputy chairman of the Commission and the cabinet members who were in the Commission. Those on the stage talked to 'them'. Then the chief ministers were given a few minutes each, strictly timed and with a buzzer to stop them, to make their remarks. When all of them had spoken, the meeting would conclude with statements from the stage. The most vocal chief ministers complained that there was neither the time, nor the intention to deliberate together.
>
> Ms. Jayalalithaa, chief minister of Tamil Nadu, one of the country's wealthier states, said in a huff that she saw no reason to come to the Planning Commission to be told how to spend her own state's money, and she walked out of a National Development Council meeting after she finished her speech and told the media outside that the format was insulting and the meeting was a waste of time.
>
> Narendra Modi, who was a three-term chief minister of Gujarat, another rich state that generated large amounts of resources internally, also had similar feelings about the Planning Commission.

After Modi came to power in 2014, the Planning Commission itself was dismantled and replaced by Niti Aayog. Aadhaar was moved to the Ministry of Information Technology. By this time, however, it was no longer the mix of start-up/sarkar that Nilekani had fashioned it into, a cross between hare and tortoise. It had evolved into a complete *sarkari* organization, captured thoroughly by bureaucrats. Bureaucrats continue to fight over their turfs, but these fights are the usual ones. They might fight to hurt, but not to kill.

The Business of Business

'It is not from the benevolence of the butcher, the brewer, or the baker that we expect our dinner, but from their regard to their own self-interest. We address ourselves not to their humanity but to their self-love, and never talk to them of our own necessities, but of their advantages.' Thus wrote Adam Smith in *An Inquiry into the Nature and Causes of the Wealth of Nations*.[12]

Having been a businessman throughout his career, Nilekani had no illusions about how his own tribe would react to Aadhaar: They will ignore it if they find no use in it, they will fight it if it threatens their business, and they will embrace it if it helps them in their business.

They might, however, not always be right in their assessment of the impact of an innovation. In *The Master Switch: The Rise and Fall of Information Empires*, Tim Wu (best known for coining the term 'net neutrality') tells the story of how AT&T invented tape-recording technology in the 1930s, but went out of its way to kill its own baby. (The technology eventually came into commercial use in the 1990s, missing out a good 50–60 years.)

Wu speculates:[13]

> But why would company management bury such an important and commercially valuable discovery? What were they afraid of? The answer, rather surreal, is evident in the corporate memoranda, also unearthed by Clark [Mark Clark, a historian who studied Bell Labs], imposing the research ban. AT&T firmly believed that the answering machine, and its magnetic tapes, would lead the public to abandon the telephone.
>
> More precisely, in Bell's imagination, the very knowledge that it was possible to record a conversation would 'greatly restrict the use of the telephone,' with catastrophic consequences for its business. Businessmen, for instance, the theory supposed, might fear the potential use of a recorded conversation to undo a written contract. Tape recorders would also inhibit discussing obscene or ethically dubious matters. In sum, the very possibility of magnetic recording, it was feared, would 'change the whole nature of telephone conversations' and 'render the telephone much less satisfactory and useful' in the vast majority of cases in which it is employed.

But in the first big battle that Nilekani had to fight with businesses, the implications were by no means ambiguous. For smartcard makers, it was as clear as seeing a delicious cheesecake slipping out of their hands.

Initially Nilekani and his team were planning to give out smartcards for every Indian. They were thinking of a model in which the government would essentially tie up with registrars—institutions that have dealt with a large number of people—such as, say Life Insurance Corporation of India, and issue co-branded cards that also displayed the unique number.

In the early days, Srikant Nadhamuni had spent a good part of his time thinking about the design of the card and the mechanism of rolling it out. But a meeting in Bengaluru in 2009 significantly changed the very nature of the product.

The meeting in Bengaluru included Abhijit Banerjee, an MIT professor who had studied 'poor economics',[14] also the name of the book he co-authored with Esther Duflo, who was also present, as well as Nachiket Mor. At ICICI—both at the bank and the foundation—Mor had taken great interest in inclusive finance. He was involved with Fino, which had experience in dealing with smartcards for the bottom-of-the-pyramid market. Mor asked a crucial question: Why were they thinking of a card? Why not just a number, which could be authenticated with biometrics?

The idea immediately struck a chord with Nilekani and his core team, in part because going for just a number meant huge savings in cost. Cards were more expensive to produce and they had to be retired every five or six years, which also meant a recurring cost.

Those who worked with the poor immediately grasped the significance. 'It might not be intuitive to those who have a bunch of credit cards within their reach all the time, but if you have never carried a wallet in your life, it can be inconvenient,' a social worker told us at that time. His NGO was trying to store immunization and other data for children on paper and smartcards, and constantly faced difficulties because people had damaged or lost their papers. He said they considered integrating a chip into a talisman—like an amulet which is quite commonly worn by traditional Indians, especially in rural areas—but for some reason it didn't work out. A number—not secret, but private, like an email ID or phone number—and biometrics for authentication seemed to offer just the right balance.

Once they decided to ditch the idea of the card, it sent a shockwave among smartcard players such as MasterCard and

Visa. 'It was as if someone pulled the rug from under their feet,' a person involved with the project said. The echoes of that fall, of the lost business in billions of dollars, can be heard till today, sometimes expressed as arguments against biometrics.

Nilekani has often stressed that the inspiration for Aadhaar has been GPS and the Internet. They were built by the US government for its own purposes. Their usefulness to people dramatically went up when the private sector was allowed to innovate on those platforms. So it will be with Aadhaar.

However, UIDAI's reliance on the private sector and businesses started earlier. It depended on the private sector to build technology, for enrolments, and to roll out the first applications of Aadhaar. Eventually, the forces of free market would lead to wider adoption by the industry.

Banks tend to be the first adopters of information technology because they primarily deal with data. The first widespread use of computers was driven by the banking sector. The banking, financial services, and insurance sector, or BFSI, was the biggest client for India's outsourcing industry.

Their appetite for information technology solutions, however, was subject to two big factors. One was regulators. Regulations can work in two ways—a lot of investments in business intelligence and risk are driven by regulatory requirement. At the same time, the adoption of innovative products driven by technology can be constrained by regulators. Nilekani's work at Infosys, and his interactions with people from the banking sector, such as Nachiket Mor, had already convinced him of how useful Aadhaar and related platforms could be for banks. To his pleasant surprise, one of the first calls came from Usha Thorat, a high-ranking official in the RBI. In her call, Thorat spoke about how useful identity would be for the banking sector, and that the RBI was keen to work with UIDAI. Nilekani had support from others in RBI's leadership team then and later, including RBI governor D. Subbarao and his successor Raghuram Rajan.

If regulators were one factor, the other was the investments that banks had already made in the existing system. It was not just about technology products, but also the entire process that was built around the system. There was resistance at several levels. Banks were already specializing in authentication (through signatures, etc) and they were not too open to outsourcing that to UIDAI. But it was not just about a part of it getting outsourced. Direct benefit transfer was fundamentally changing the way things worked. In the existing scheme of things, the entire subsidy transfer for cooking gas was handled through the government-owned State Bank of India (SBI), the largest bank in India. In effect its job was to handle four large cheques from the government to oil marketing companies. In the new scheme of things, it is about doing millions of small fund transfers directly to customers—to bank accounts that might not even be with SBI.

Nilekani's method didn't differ from the way he used to deal with his clients when he was CEO at Infosys. Now, he held the rank of cabinet minister, but he didn't give too much importance to the protocols. He understood that ultimately businesses looked to protect their interests and their revenue streams.

Watching his presentations on banking, financial sector, media, etc., can give an idea of how he might have approached the business leaders at that time. Typically, he talks about the big trends—the business, economic, social, and technological—that are shifting the landscape. These trends are like the force of gravity, and all one can do is accept they are inevitable and change. And then he takes them through the solutions and tries to convince them that they need not be swept away by these big forces, that there is a way to adapt and thrive.

Law and Politics

> Wendy: Do you care about your relationships at Axe Cap? How are those relationships?
>
> Taylor: Good.
>
> Wendy: Why do you think that is?
>
> Taylor: You know what? The culture is set from the top. If Axe likes me, other people do. Or pretend to.
>
> Wendy: Why do you think he likes you?
>
> Taylor: One of the two reasons why anyone likes anyone else. Either they recognize a part of themselves. Or they see something they can use. In this case I imagine it's both.
>
> *Billions* (2017), Season 2, Episode 8, 'The Kingmaker'

In his initial days in the government, Nilekani had a halo. After all, he hadn't lobbied for this job; Rahul Gandhi had personally invited him. He would meet Manmohan Singh every month to brief him on the topic. While Singh was the prime minister, the most powerful person in the ruling party—after the Gandhis—was Pranab Mukherjee. He had assured Nilekani that he would watch his back so that Nilekani could go about his work. However, in July 2012, Mukherjee left the government to be sworn in as the president of India, and UIDAI lost a champion. Fortunately, by that time, Chidambaram had moved from the

Home ministry to Finance and had started speaking the language of his new bureau. The winds were constantly changing.

Through all these changes though, Nilekani was hoping that the government would pass the Aadhaar Act in parliament. Almost all large-scale government programmes are backed by a law passed by parliament. UIDAI was backed only by an executive order. Without a law backing it, the project mostly relied on his ability to explain the benefits of the project to different states to set up memorandums of understanding (MoUs).

In August 2010, UIDAI presented a draft National Identification Authority Bill, kicking up a range of debates. The Home ministry was worried that the bill gave too much space to UIDAI, infringing upon its own territory. Civil society activists were worried that it would lead to a disruption of government welfare programmes. There were concerns about privacy issues in a country that had no privacy or data protection law. India was getting digitized thanks to mobile and internet penetration, and with or without Aadhaar, there would have been a need for these laws. However, Aadhaar had now become the rallying point.

In December that year, the cabinet approved the bill, passed it in the Rajya Sabha and formed a parliamentary committee headed by Yashwant Sinha to review it. A year later, after meeting with various groups, getting inputs on questions ranging from purpose, cost, impact, issues with NPR, it rejected the bill, sending it back to the government. Gurudas Dasgupta, Communist Party of India (CPI) leader and a member of the committee, told a newspaper that they felt there was no need for such a project because there were alternatives to identification. Even though the committee had 11 members from the Congress party, only three filed dissent notes.[15]

UPA 2 did not come back with a revised bill. Meanwhile enrolments continued. When the BJP-led NDA government came to power in 2014, it took the case up, drafted the law, and was looking to pass it through both houses of parliament. It had a majority in the Lok Sabha, but not in the Rajya Sabha. During this phase, Nilekani tried to convince Rahul Gandhi that since Aadhaar was an initiative of his party, they should consider openly supporting this bill, and make it a bipartisan initiative. Given the muted support from some Congress party leaders when they were in power, Rahul Gandhi let it go. Meanwhile, the BJP was so unsure of any support from the Congress that it passed it as a money bill, which meant that the Rajya Sabha could offer suggestions, but not vote for or against it. It continues to be a thorn. Congress party leader and former minister of Rural Development Jairam Ramesh, who was once Nilekani's classmate at IIT, has filed a case in the Supreme Court on the legality of passing it as a money bill. Chidambaram is arguing the case in the court.

Vested Interests

One of the criticisms of Aadhaar in its early days was that it was just an identification. It did nothing else. It didn't tell you if you were eligible to drive or to travel abroad or for government subsidies. People who did not have any ID—or at least none that were widely accepted or trusted—instinctively grasped the idea, getting some comfort from the fact that it was given by the government, and there was no enrolment fee. But many—especially those who already had multiple identity documents—saw no need for yet another identity. In any case, it was important for UIDAI to show that it had uses, that it could benefit the citizens, and make the delivery of public goods more efficient.

The reason the government launched the project was not merely to give every citizen an identity; it also wanted to use it to deliver benefits. In the budget speech of 2011–12, Mukherjee announced that he was setting up a task force under Nilekani to explore direct benefits transfer (DBT), and in June, Nilekani submitted an interim report that looked into providing subsidies on cooking gas, kerosene, and fertilizers.

He knew it was not going to be an easy matter. The report said:[16]

> The social programs of India are complex systems with millions of participants that have evolved over the last few decades. Hundreds of millions of beneficiaries depend upon these programs for basic sustenance. And neither is technology a panacea. Eventual success will hinge upon political will, good governance, incentive-compatible solution design, judicious use of technology, a structured transition plan, meticulous project management, effective supervision, audit and execution.

The task force's proposal for cooking gas was simple. Instead of selling cylinders to customers at a subsidized price, the oil marketing companies would sell them at market price, while the government, instead of paying the subsidy in bulk to oil marketing companies, would directly transfer the subsidies to the beneficiaries. They just had to link their Aadhaar number to a bank account for the money to be transferred there. A limit was set on the number of cylinders a household could claim the subsidy on—at nine per year. This was met with opposition from the Liquefied Petroleum Gas (LPG) distributors' lobby.

The move made it difficult for the middlemen to divert subsidized cylinders to commercial establishments.

Meanwhile, the Congress was facing heat from voters in Delhi (where Sheila Dixit went out of office after being elected chief minister for three terms) and Rajasthan (where the number of its seats came down drastically). Congress politicians, including Jairam Ramesh, attributed the losses in part to DBT—and convinced the leadership to press the pause button on the DBT programme.

The message was clear. Politicians tend to be swayed by pressure groups and what they believe to be the mood of the public. They don't hold on to any specific position. But this also turned in favour of UIDAI.

If one goes by the rhetoric during the general elections held in April–May 2014, BJP was opposed to Aadhaar. In fact, a reason why Congress developed cold feet on DBT was not just because its own party members were expressing doubts, but also because these views were amplified by the Opposition. Many who had worked closely on the project were keeping their fingers crossed during and after the elections. Even those within the BJP government were not sure what the way forward was.

Nilekani knew that nothing can be taken for granted. The fact that over half the population of the country had an Aadhaar was no guarantee. As Chidambaram told us during an interview in 2017, if a government decided to scrap Aadhaar, there is nothing that can stop it. But there were already hints that there would be movement on Aadhaar. Ram Sevak Sharma, after his tenure at UIDAI, had gone to Jharkhand as the state's chief secretary and had implemented an Aadhaar-based attendance system for the state government. Prime Minister Modi, who was known to be an efficient administrator, found that application interesting. Later, when Sharma, as secretary of IT, made a presentation on Aadhaar to the leadership of the new government, it again made an impression on Modi.

Nilekani, who had met Modi when he was Gujarat's chief minister and spent over an hour talking to him, sought an appointment. It was duly granted on 1 July 2014. Modi was interested in Aadhaar's applications in public distribution. He had questions about security. But he was already convinced for most part. Maybe, the meeting helped to give the final nudge, Nilekani said.

Over the next few months, the Modi government passed the Aadhaar Act (even though as a money bill), made sure it would not merge with the Home ministry's database, and increased the number of Aadhaar-based applications at breakneck speed. The scope and speed would create its own set of problems.

When Ideologies Meet Ideas

Dealing with activists, however, turned out to be different. In the early stages, Nilekani believed that they would be powerful allies because the purpose was common. Activists wanted to solve problems, and UIDAI was promising to do the same, using technology. Their negative response took him by surprise.

'Nandan kind of approached them the way he would approach a business leader. Businessmen want to solve problems, or want to get their problems solved. And, typically, when they encountered someone who offered to help, they listened. They were pragmatic. Nandan somewhat naively believed the same approach would work with activists as well. But, they didn't appreciate a stranger coming in their midst and telling them that he has a better way to solve the same problems that they have spent their lifetimes trying to solve,' a person who had seen Nilekani at that time said.

'At an event organized at the National Law School in Bengaluru, Nandan, along with his team, faced a room full of activists,' he added. 'The activists' response was visceral. Nandan seemed somewhat taken aback, and his response was to give an elaborate answer to every question. His team was sitting right next to him. I don't know if he was being protective of his team members, but he hardly let them speak. Then there was a tea break. His wife Rohini was in the audience and must have sensed the mood. She advised him to be a bit more relaxed. In the session after the tea break, Nandan held himself back and let his team take questions.'

It was not just about an event in Bengaluru. At that time, activists such as Aruna Roy and Nikhil Dey, and academic-activists like Jean Drèze, wielded enormous influence on the government. They were behind some of the most far-reaching social sector legislations that the UPA had put in place, including Right to Information, food security, Right to Education, and the Rural Employment Guarantee scheme. Roy and Drèze were members of the National Advisory Council, set up by the first UPA government to advise the prime minister on the government's social agenda. Both were lobbying with Sonia Gandhi, who was the council's chairperson, that the project should be reviewed by the council. Nilekani knew that would have killed the project. He sought a meeting with Sonia Gandhi, and took with him photographs of people standing in queues to enrol for Aadhaar. They were doing this with no law in place, with no government programme tied to it. Nilekani asked her, Aadhaar is popular among people; why would you want to stop it?

Eventually, the composition of the National Advisory Council changed, and most of the new members had a positive view on Aadhaar. The intensity of some of the initial criticism—that questioned whether it could be executed well, and whether all the citizens would get an Aadhaar—had gone down, thanks to the speed of enrolment. During one council meeting, Roy's was the lone voice opposing Aadhaar.

With the new government led by Modi, activists have far less say in running the government, leaving them only the media and the courts to highlight their concerns.

For Nilekani, dealing with the activists was a lesson. He learnt that they were primarily driven by a narrowly defined ideology. Sometimes, that created problems among two different sets of activists on some matters of principle. For example, Right to Information activists were insisting on proactive transparency in government expenditure on subsidies. This would help the public verify if the subsidies are going to the right people. But then, such transparency was anathema to privacy activists, because they believe that the individual's data should be kept private.

Nilekani also had to face some kind of resistance at home, for his wife Rohini is a self-described Left liberal. 'I was very frightened and opposed to Aadhaar from the moment I heard about it, for nearly one-and-a-half years. Because I am a knee-jerk liberal. And knee-jerk liberals came immediately to the fray thinking that it will become a tool for surveillance. I am suspicious of the state anyway—and that has nothing to do with Aadhaar. And many of us on the Left are. We feel one of the big lessons of history in the last century shows that the state can get very aggressive. So, we had a lot of arguments. Loud ones. I was loud and aggressive. After one-and-a-half years of that, I began to see it more in the way Nandan saw it. I saw the big picture. I could see how transformational it can be, if it's done right. So, I began to think more deeply on it. We have 500 million people who don't have access to basic services. Aadhaar is one of the things—not the only thing—to help them. Modern technologies are converging to allow such things to happen,' she told us.

Nilekani might have found peace at home, but when the Modi government decided to step on the gas on Aadhaar linkages, it also gave momentum to civil activism, not just from the NGOs on the ground, but also from technology activists. When UIDAI was fighting some of the toughest battles, Nilekani was no longer a part of the organization. He was not even a part of the government. But he was associated so closely with the project that he was very much at the centre of those battles too.

Vikram's Answer

Vikram, in the story we started with, would have had no difficulty in giving the answer. That's in part because house numbers are so much a part of the existing system that arguments against it actually sound ridiculous. Some of them, even if they had some merit at one point of time, have found their fixes.

With Aadhaar, we're right in the middle of its roll-out. The same arguments that sound weird in the made-up city in the story are very valid in the context of Aadhaar, because we have not figured out good answers to the questions of exclusion, fraud, and intrusion. Not just that, we're still in the process of figuring out ways even to use Aadhaar.

Notes

1. John W. Kingdon, *Agendas, Alternatives, and Public Policies* (New York: Pearson, 2013, 2nd edition).
2. Sun Tsu, *The Art of War*, trans. James Clavell (Sweden: Chiron Academic Press, 2015).
3. PTI, 'Govt Serious on Multi-purpose National ID: Qureshi', *DNA*, 3 June 2009, http://www.dnaindia.com/india/report-govt-serious-on-multi-purpose-national-id-qureshi-1261474, viewed on 2 July 2018.
4. Indian Kanoon, 'The Citizenship (Registration of Citizens and Issue of National Identity Cards) Rules, 2003', https://indiankanoon.org/doc/199236652/, viewed on 2 July 2018.
5. Usha Ramanathan, 'The Function Creep that Is Aadhaar', *The Wire*, 25 April 2017, https://thewire.in/featured/aadhaar-function-creep-uid, viewed on 2 July 2018.
6. Malini Bhupta, 'RIL Feud: Anil, Mukesh Ambani Get Ready for Long-Drawn Battle, Kokilaben Holds Key to Truce', *India Today*, 13 December 2004, https://www.indiatoday.in/magazine/economy/story/20041213-ril-feud-anil-mukesh-ambani-get-ready-for-long-drawn-battle-kokilaben-holds-key-to-truce-788931-2004-12-13, viewed on 2 July 2018.
7. Paul Ormerod, *Why Most Things Fail* (London: Faber and Faber, 2010).
8. Amruta Byatnal, 'Tembhli Becomes First Aadhar Village in India', *The Hindu*, 29 September 2010, www.thehindu.com/news/national/Tembhli-becomes-first-Aadhar-village-in-India/article13673162.ece, viewed on 2 July 2018.

9. Shekhar Gupta, 'National Interest: The Officer Raj', *Indian Express*, 15 October 2011, http://indianexpress.com/article/opinion/columns/national-interest-the-officer-raj/, viewed on 2 July 2018.
10. Gupta, 'National Interest'.
11. Arun Maira, *An Upstart in Government* (New Delhi: Rupa, 2015).
12. Adam Smith, *An Inquiry into the Nature and Causes of the Wealth of Nations*, https://www.econlib.org/library/Smith/smWN.html).
13. Tim Wu, *The Master Switch: The Rise and Fall of Information Empires* (London: Atlantic Books, 2010).
14. Abhijit Banerjee, *Poor Economics* (New Delhi: Random House India, 2011).
15. *The Hindu*, 'CPI Leader Gurudas Dasgupta Challenges Party Line', 4 April 2012, https://www.thehindu.com/todays-paper/tp-national/tp-newdelhi/cpi-leader-gurudas-dasgupta-challenges-party-line/article3278863.ece, viewed on 2 July 2018.
16. Ministry of Petroleum and Natural Gas, 'Report of the Task Force on an IT Strategy for PDS and an Implementable Solution for the Direct Transfer of Subsidy for Food and Kerosene', 2011, petroleum.nic.in/sites/default/files/pdsreport.pdf, viewed on 2 July 2018.

3

The Platform Paradox

THE RESIDENTS OF A RELATIVELY CALM PART OF Frazer Town, a locality in central Bengaluru known for its schools and bakeries, noticed that they had a new visitor—a man who was probably in his twenties or thirties. He seemed to have come from nowhere. He was about 5 feet tall, sported an unruly mop of hair; his clothes were dirty and torn, and he carried a sack on his back. He moved from one road to another, rummaging the garbage in street corners, next to transformers, and near the compound walls of unoccupied buildings.

When he wasn't moving around, he'd settle himself in front of an old, dilapidated house next to a heap of waste, talking to himself. Sometimes, when passers-by gazed at him with curiosity, he'd reach for a stick and wave it threateningly. For the kids (who played cricket, football and other games as the season demanded), he had a bunch of tiny stones by him. They let him be.

Sometimes, a lady in the neighbourhood would leave something to eat in a polythene bag next to him. He would look uninterested, talking to himself, staring at the trees. A few minutes after, when he was sure she had left, he would pick up the bag and finish off its contents.

Months passed.

One morning someone saw him lying sideways in the foetal position, an outsized embryo, back curved, arms and legs bent towards his torso, motionless, some metres away from his usual resting place. He was dead. A few people gathered around him, someone pulled out his mobile to call the police. About 30 minutes later, a couple of policemen arrived, and made a few calls to get a municipal corporation van to take away his body. He was gone as he came—no identity, no name, no proof of existence, only a record of his death.

A Widespread Phenomenon

Across the country, thousands die this way. A survey by Delhi police in 2004 found that most of these unclaimed bodies were those of 'rickshaw pullers, rag pickers, beggars, porters, hawkers, construction and domestic workers'.[1] Zonal Integrated Police Network, an initiative that followed the survey and included—besides Delhi—Haryana, Punjab, Chandigarh, Uttar Pradesh, Rajasthan, Uttarakhand, and Himachal Pradesh, shows over 70,000 records of unidentified dead bodies as of July 2018.[2] In Chennai, where similar issues exist, S. Sreedhar, an executive at a finance firm, was so distraught by the indignity of such deaths that he formed a trust to give a decent burial to such people.

In many ways, the anonymity in death is just a continuation of anonymity in life. One need not go around too far in any part of the country before encountering such people. They are homeless, living on the roads, behind temple walls, on and under bridges, or by the banks of waterways now carrying sludge. That they survive in harsh conditions has little to do with government.

They have learnt to survive, sometimes using their own resources and sometimes by the benevolence of others—such as the kind shown by Narayanan Krishnan. Krishnan, a chef at Taj Hotels with a promising career, was visiting his home city Madurai in 2002. He witnessed an old man eating his own excreta because he couldn't find any food. It disturbed him enough to quit his job and start feeding the homeless and differently abled. His organization, the Akshaya Trust, feeds over 400 people every day.[3]

Life is difficult not just for those in extreme poverty. It's difficult for many at the bottom of the pyramid (a term popularized by management guru C.K. Prahalad to indicate people slightly above those in absolute poverty). Many in this group, who have jobs and have accumulated some physical assets, struggle. They cannot leverage these assets because they don't have documentary proofs. As Gurcharan Das writes in his book *India Grows at Night*,[4] 'The poor, thus, have houses but no titles; crops but no deeds; and business without licenses. The informal economy is people's spontaneous response to state failure.'

Pockets of Excellence

Over the years since Independence, various governments have tried to address this segment through a number of poverty alleviation programmes: old-age pensions, employment guarantee schemes, PDS, midday meal programmes, and so on.

Some of these programmes are well run, and have received widespread admiration. Tamil Nadu, for example, has been running a midday meal programme for children in government schools since the 1960s, a scheme that was expanded and eventually adopted nationwide. It has had second-order and long-term impacts. A study by Farzana Afridi of Syracuse University, New York, and the Delhi School of Economics found that it increased school attendance of girls by 12%.[5] Later, under Tamil Nadu chief minister J. Jayalalithaa, the state government launched a chain of low-cost restaurants providing basic food for as low as Rs 10 per plate. There are indications that other states might follow suit. Karnataka, for instance, launched its own version in 2017.

Another example is polio eradication. India had 50,000 cases of polio in 1994.[6] The workers on ground had to deal with physical stress, logistical complexities, and even rumours that the vaccination programme was a ploy to make citizens infertile. In 2014, the World Health Organization declared the country polio-free. 'We as a civilization have few things we can accomplish of genuinely lasting significance for mankind: we have built no pyramids, no Great Walls to stand for thousands of years. It is, instead, through medicine that we may create our enduring monument. The eradication of smallpox and now, perhaps, polio will stand as our pyramids,' author, physician, and health policy expert Atul Gawande wrote in the *New Yorker*.[7]

Pipes and Platforms

While examples such as these show that the government is capable of achieving excellence, the overall performance has been disappointing so far, especially with respect to the poor. While the government spends $60 billion every year in terms of targeted subsidies, it's not clear how much of it reaches the poor. Former Prime Minister Rajiv Gandhi observed in 1985 that for every rupee the government spends, only 15 paise reaches the intended beneficiaries.[8] A Planning Commission report published in 2005 said that more than a third of the grain intended for poor households was diverted to others, and less than half of subsidized grains reached targeted beneficiaries.[9] Speaking at a seminar in Delhi in 2009, Planning Commission Deputy Chairman Montek Singh Ahluwalia echoed what Rajiv Gandhi said 25 years earlier—little had changed since then—and suggested that 1% of any subsidy scheme costs go towards monitoring and evaluation.[10]

When such statements are made, we tend to place great emphasis on government inefficiency and wastage of resources, and ignore the impact it has on the poor. They end up not getting what they are entitled to, and are harmed twice, first by poverty, and then by elitist complaints about subsidies and income distribution. 'Our provision of public goods is unfortunately biased against access by the poor,' Raghuram Rajan, former governor of RBI, said in a 2014 speech.[11]

Fair price shops, schools, police stations, public hospitals—all these facilities were built for the poor, but fail them frequently. And where they are available, they often are of such poor quality that even the poor try to avoid using them. When the poor get money, they send their children to private schools, and not government schools, because the quality is better. When they can afford it, they go to private doctors rather than primary health centres. Sales representatives of Unilever, trying to sell Lifebuoy soaps and Lipton tea to village shops made sales forecasts based on the queues outside public health centres. 'When the crowds start to swell at the public health centre, business dries up,' a sales representative explained.

India has failed to build state capacity to deliver public goods (a point Raghuram Rajan raised in his Kosambi Ideas Festival speech in 2015).

What is common to many of these programmes is that they are *solutions*. Sangeet Paul Choudary in his books *Platform Scale* and *Platform Revolution* contrasts platforms with pipes.[12] In the pipe model, value gets created at one end (by the producers) and it gets consumed at the other end (by the users). Hollywood produces movies at one end, distributes them through cinema halls, television channels, DVDs, etc, and at the other end of this supply chain, people watch the movies. A platform, on the other hand, just attracts producers and consumers to it, and enables transactions between them. YouTube lets producers post their videos and users watch them. Value is created on the platform.

Platforms can scale up (the growth of Uber, Airbnb, YouTube are the most popular examples). Platforms can cater to a range of services (there are platforms for education, health, and entertainment). Platforms can nurture diversity (the range of travel options available on Ola is wide). Platforms can gain momentum very fast (one reason why venture capitalists are attracted to platform businesses). Platforms can dramatically change the industries they operate in (one reason why they meet with opposition from the incumbents).

Aadhaar was built as a platform. Nilekani said that when they were building Aadhaar, they took GPS and the Internet as inspirations. Both GPS and the Internet were US government projects, and both had an hourglass architecture, allowing innovations above and below them. In a similar manner, Nilekani and team envisioned Aadhaar at the stem of the hourglass, with innovations in devices (such as scanners) happening below the stem and innovations in applications (say a PDS application using Aadhaar) happening above the stem.

Aadhaar just needed to ensure that different applications could access its functionality in a safe and secure manner.

An Outsized Technological Infrastructure

This approach has received global attention and admiration by many who saw that the government looked at it as a platform, as digital infrastructure. Investor Raoul Pal, a former hedge fund manager and CEO of Real Vision Group, said in a widely read note: 'India has built the world's first national digital infrastructure, leaping at least two generations of financial technologies and has built something as important as the railroad was to the UK or the interstate highways were to the U.S.'[13]

Some others saw its potential to have a huge impact. A UN report published in 2016 said it had 'tremendous potential to foster inclusion by giving all people, including the poorest and most marginalised, an official identity'.[14] Jim Yong Kim, president of the World Bank, once said about Aadhaar: 'This could be the greatest poverty killer app we've ever seen.'[15]

The outspoken Paul Romer, chief economist of the World Bank until January 2018, told Bloomberg: 'The system in India is the most sophisticated that I've seen. It's the basis for all kinds of connections that involve things like financial transactions. It could be good for the world if this became widely adopted.'[16]

However, for the team that built Aadhaar, the most satisfying vote of confidence came as the project was getting rolled out. Millions of poor stood in queues to enrol themselves, long before it was linked to any programme. Shankar Maruwada said, 'Speaking to them we got a sense that they looked at identity as an asset. In one village, when they asked a man who already had a couple of identity documents why he wanted one more, he replied, "If you have two buffaloes, and the government gives one more, will you say no?"'

At the same time, Aadhaar had also attracted a number of critics—some of them demanded that Aadhaar and Aadhaar-based applications should be improved, and some called for suspension or outright destruction of the programme, citing concerns related to exclusion, security, privacy, and misuse.

Interestingly, some of these concerns about Aadhaar arise from the same source that gives it strength: the way it is designed.

One, it's a Lego block, one piece in a larger system, not a solution by itself, but something that can accelerate a solution.

Two, it has an 'unbundled' identity—and provides a way for residents to authenticate themselves. It does only this job, and nothing else.

Three, it uses biometrics for deduplication, and uses a federated model for authentication, including biometrics, OTP, PIN, etc.

Four, as a system, it depends on incentives, and getting them right is important for it to evolve and to succeed.

Finally, all said and done, it's a government project, and so its methods don't always go well with the incentives on which its continued evolution depends.

A Lego Block

On a Monday morning in June 2017, Pramod Varma, the chief architect of Aadhaar, tried to convince a group of social workers, development officials, international diplomats, and technologists assembled at the UN headquarters in New York that Aadhaar does nothing. That is, Aadhaar does nothing by itself—and that is its strength.

They were assembled at the UN headquarters for the 'Platform for Change' Summit. The organizers, ID2020, an alliance between governments, NGOs, and businesses, were hoping to 'spread awareness of the challenges faced by those living without recognized identification, explore the role of identity as a platform for social and economic opportunity, and move towards development of a coordinated roadmap for action'.[17]

Aadhaar, Varma told the gathering, meant 'foundation' in many Indian languages, and is an identity project. It was that simple.

> Aadhaar does not give any entitlement. Zero. It does not establish citizenship. It does not guarantee food. It does not guarantee a job. It does not guarantee banking services. It doesn't guarantee anything. All it does is provide you a unique identity.[18]

A large number of Varma's audience in the summit were in the development sector; they had worked in diverse settings and were aware of the lack of government capacity in many emerging economies. In fact, those who were working in forced migrations and human settlements were exposed to the worst failings of the state.

Varma explained to his audience:[19]

> We looked at the problem very differently. We said, we should not be solving subsidy issues, we should not be solving health issues, we should not be solving education issues in a silo. To say that we needed to solve these issues would mean that we assume we know the solution. It's almost always a bad assumption that you know the solution to the problem—especially in a large and diverse country like ours. You don't know the solution to a problem. Most of us don't know.

Over several conversations we had with Aadhaar's founding team, one thing became clear. Instead of building a solution, they were asking, why not improve the government capacity to come up with better solutions?

As Varma had said,[20] 'The way to solve India's hard problems, wicked problems, was not by building solutions, but by building Lego blocks that could be used by people on the ground, closer to the problem, to assemble, build solutions.'

The Aadhaar Man on the Ground

One person who was right on the ground, close to the problem, and was raring to build solutions using technology was A. Babu, a 2003-batch IAS officer from Kerala. Under him, Krishna district in Andhra Pradesh, where he served as the collector from 2015 to 2017, became an epitome of Aadhaar-based public service delivery.

Babu is a stocky man who walks with a sense of urgency, enthusiasm, confidence, a ready smile plastered on his face. One of his colleagues said initially he had a suspicion that he never slept. He had once seen Babu off after a tiring tour at 3 in the morning. Later, when he woke up mid-morning to check his Telegram app, he saw that his boss had been busy replying to messages, sending instructions to his other colleagues right from 5 a.m., as usual. He was so enthusiastic about Aadhaar that he used to be referred to as Aadhaar Babu (even though the 'A' in his name stands for Ahmad).

During the days when he was trying out different Aadhaar-based technology solutions, Babu used to carry around three or four mobile phone models—all different from the ones his close team members carried. When technology companies, start-ups, tech enthusiasts came to him with their solutions, he wasted no time in testing them out on these phones.

The results of his initiatives can be seen at the ration shops spread across the district.

At Chinnampetta, one afternoon, a ration shop on the side of a small dusty road had just two customers. 'It's usually crowded in the first week, and then the flow drops. We sell all goods by the tenth of every month,' the ration shop owner said. An old man walked in with a cloth bag. The shopkeeper asked for his Aadhaar number, entered it in an ePOS (electronic Point of Sale) terminal, and asked him to show his left ring finger (the ePOS terminal indicates the best finger to scan). On the small screen, his entitlements showed up. The old man wanted to buy 10 kg of rice. The shopkeeper weighed the required quantity on a weighing machine connected to the ePOS terminal. The sale was complete. The machine then connected to the old man's bank account for the money transfer. 'Earlier, people used to look at us with suspicion, and wouldn't believe if I said I didn't have change. Now it's digital—money gets transferred online,' he said.

One morning, M. Mahalakshmi Amma and her daughter, P. Annapurna, walked into Bhavanipuram Fair Price Shop, not to buy rations, but to withdraw some money from their bank accounts. They are both old and have lost their family members over the years, and now take care of each other. Their monthly pension is deposited into their accounts. The ration shop owners were also armed with micro ATMs, a human-operated, fingerprint-enabled POS terminal which the customers could use to withdraw cash (and digitally transfer the amount to the Fair Price Shop owner). Mahalakshmi Amma said that earlier she would go to a bank branch 2 kilometres away to withdraw her pension. Now it was literally a few metres away.

For most beneficiaries, the use(fulness) of Aadhaar begins and ends in a ration shop. It's in fact just a single step in the process of buying their products. The system itself is much larger.

This point cannot be overstated or overcommunicated, because much angst around Aadhaar comes from the idea that it is a solution by itself. A joke in bureaucratic circles captures this well: If you have a hammer in hand, every problem looks like a nail. If you have Aadhaar, every problem looks like a fingerprint.

Many understand the impact of Aadhaar to be so. Reetika Khera, who teaches at IIT Delhi, describes Aadhaar as 'a solution looking for a problem'. Usha Ramanathan, a legal researcher also based in Delhi, looked at this quality with much suspicion. 'Nandan is marketing Aadhaar depending on who the audience are. When he talks to school administrators, Aadhaar is projected as a solution to solve attendance problems.

When he is talking to finance people, he projects Aadhaar as if it would solve their KYC woes. So, it is all things to all people. It's basically marketing,' she said.

Aadhaar could never have lived up to such expectations because it was designed as an enabler.

In his commencement address to Caltech students in 1974, American theoretical physicist Richard Feynman gave an interesting metaphor to think about science in the right way. He said:[21]

> In the South Seas there is a Cargo Cult of people. During the war they saw airplanes land with lots of good materials, and they want the same thing to happen now. So they've arranged to make things like runways, to put fires along the sides of the runways, to make a wooden hut for a man to sit in, with two wooden pieces on his head like headphones and bars of bamboo sticking out like antennas—he's the controller—and they wait for the airplanes to land. They're doing everything right. The form is perfect. It looks exactly the way it looked before. But it doesn't work. No airplanes land. So I call these things Cargo Cult Science, because they follow all the apparent precepts and forms of scientific investigation, but they're missing something essential, because the planes don't land.

Feynman's metaphor can well apply to thinking about technology and society. Broadly, the South Sea islanders mistook a form, a front end, for the system, because the front end was what they saw. What brought them the food were the things they couldn't see. Robert Pirsig, author of *Zen and the Art of Motorcycle Maintenance*,[22] divides thinking into two types: romantic and classical. The romantic way is to look at the surface; the classical way is to look at the underlying structures.

Babu, who in May 2017 moved on to assume a bunch of responsibilities including CEO, Real Time Governance, vice chairman and managing director, Andhra Pradesh Industrial Infrastructure Corporation, and managing director, Andhra Pradesh State FiberNet, had started thinking about using Aadhaar long before he became the collector of Krishna district. He put in place a pilot in East Godavari district, when he was a joint collector there between 2011 and 2013. The experimentation continued when he moved to Adilabad. (One of his colleagues considers it a punishment post, because the implementation of Aadhaar and the resulting transparency had hurt some vested interests.) Before long, Chandrababu Naidu, the technology-smitten, forward-looking chief minister of Andhra Pradesh, noticed him, and brought him to Krishna district to implement on a larger scale what he had started in East Godavari district.

Babu's work in the back end demanded back-breaking physical work, technology implementation, and also taking on and responding to different sets of stakeholders who stood to lose out by the changes.

When he was working on the pilot in East Godavari district, one of the first things Babu ensured was that everyone had an Aadhaar number. It was early days, and the project was being rolled out. What helped Babu was also a decision made by the UIDAI team to not spread the enrolments thin, but to ensure that there were pockets which had 100% enrolment precisely for this reason. Babu also wanted to make sure that the beneficiaries who couldn't link their Aadhaar to the PDS database were not counted as 'ghosts'. He followed it up with a door-to-door survey. When he came to Krishna district he had a good idea about the issues on the ground, and so he put in place good exception-handling mechanisms: if the fingerprints didn't work three times, they would use an iris scanner; if that didn't work, they would use mobile OTPs. If all these failed, the village revenue officer would okay the transactions. The idea behind putting such safety nets was that while biometric authentication might have drawbacks, the system itself should not exclude anyone. (For example, if it takes four attempts to get authenticated via fingerprint, its failure rate is 25%. Yet, the system should have back-ups—other biometrics, mobile or manual overrides—to ensure the system itself does not say 'no', when it should say 'yes'.)

It was not the front end that was automated, but the entire supply chain—right from procurement, to transport, to storage in godowns, to delivery to the ration shops. A supply-chain management dashboard lets *mandal*-level stock officers (*mandal* is the intermediate level in the rural local administration structure) keep track of the real-time movement of materials, using GPS. At any point in time, the system knows the quantity of goods that are in the godown, in transit, with the ration shop, and claimed by beneficiaries, reducing the chances of leakage anywhere in the system. 'The verification at the last mile, when we ask the beneficiaries to authenticate themselves, is not because fraud happens there,' one government official told us, 'but because the leakage happens somewhere else but gets attributed to the points where there is no visibility. By bringing visibility at that point, Aadhaar helped plug leakages in the system.'

There was resistance to such changes from multiple quarters. Some fair price shop owners resisted because they could not sell the items at market price and pocket the difference.

The government not only increased their commissions to 70 paise a rupee from 20 paise a rupee, it also offered them the option of becoming banking correspondents, so they would have an additional means of legitimate income. Some of the reasons were local. In seafaring villages near Machilipatnam, fishermen staged a protest. The reason: when the leakage stopped, supply of kerosene in the black market dropped, and fishermen who depended on that for fuel were hurt.

By no means can the implementation of Aadhaar-enabled systems in Krishna district be called perfect. They made mistakes, but learnt from them and made the system better. Among the beneficiaries there was a dip in experience before it became better.

That Krishna district has gotten better over time does not guarantee that others will learn from its evolution. In part, it's because local conditions vary. It was relatively a better-off district, with per capita income 20% above national average, backed by better roads, power, and literacy rates. The collector, highly motivated, passionate by himself, had the backing of the chief minister.

Aadhaar, in effect, was just one factor that helped the system work.

Kentaro Toyama, author of *Geek Heresy: Rescuing Social Change from the Cult of Technology*,[23] and a co-founder and assistant director (2004–09) of Microsoft Research India, a computer science lab based in Bengaluru, has argued that intent and capacity are even more important than technology.

Toyama wrote in the *Atlantic*:[24] 'In project after project, the lesson was the same: information technology amplified the intent and capacity of human and institutional stakeholders, but it didn't substitute for their deficiencies.'

The experience in Krishna district in many ways confirms this view—Aadhaar is just one Lego block in a technology solution, and technology itself is only one part of the broader system as it struggles to meet its goal. The Krishna district experience also shows that a promising and a crucial piece of technology can spur action towards making the entire system better.

Aadhaar Unbundles Identity

When Nitin Pai, founder and director of public policy think tank Takshashila Institution, sold his house in Singapore before moving to India, the entire transaction took just about 15 minutes. However, when he bought a house in Bengaluru, the process stretched for a few months. The reason: in Singapore, the basic elements that go into such a commercial transaction, identities of the buyer and seller, ownership, etc., are digital.

In India, not only were they not digital, for many they were missing in any form. One of the most memorable ways in which it struck Pai was when he was talking to an activist who at one time was involved in a project to persuade people in Naxal areas to vote. It was a tough task because the people didn't want to risk their lives—the Naxalites had threatened to kill them if they dared. In one such hamlet, he was having a relaxed conversation with an elder and happened to show his voter identification card. The old man was intrigued and asked him to explain what it was. He did.

'Will I get one of these if I vote?' the villager asked.

'Yes, indeed.'

Then he went into a huddle with some of his friends, and retuned to announce that they would vote—even if it meant going against the diktat of the Naxals. What made them do it? If they got a card, they would have some documentary proof of their existence in that place.

For many Indians who have multiple proofs of identity and addresses, it is encounters such as these that highlight that millions of Indians go without an identity that is widely accepted. They might have a paper that helps them claim their benefits from the government. But, often, these papers are worthless if they stray far from the place where they are known socially.

India was facing an identity crisis: a large chunk of the population did not have an identification that was widely accepted. The ones that were available—ration cards, voter IDs, and passports—were too local or too exclusive, and in some cases too untrustworthy. Arvind Virmani, chief economic advisor between 2007 and 2009, had suggested the creation of an ID programme. Raghuram Rajan, in his report on financial reforms, *A Hundred Small Steps*,[25] had argued that a national ID was necessary for financial inclusion. In fact, Rajan and fellow economist Abhijit Banerjee were among those contacted by management professor C.K. Prahalad who was compiling a list of ideas for Prime Minister Manmohan Singh. Providing a unique identity for every Indian was one of those ideas. Nilekani himself had written about the need for an identity in his 2008 book:[26]

> Creating a national register of citizens, assigning them a unique ID, and linking them across a set of national databases, like PAN and passport, can have far reaching effects in delivering public services better and targeting services more

> accurately. … Unique identification of each citizen also ensures a basic right—the right to 'an acknowledged existence' in the country, without which much of [the] nation's poor can be nameless and ignored, and governments can draw a veil over large scale poverty and destitution.

Delivering his budget for the financial year 2011–12, then Finance minister Pranab Mukherjee made an announcement that gave the first clear indication of how the government planned to use Aadhaar: 'To ensure greater efficiency, cost effectiveness, and better delivery for both kerosene and fertilizers, the government will move towards direct transfer of cash subsidy to people living below poverty line in a phased manner.' He also announced that Nilekani would head a task force to work out the modalities.

Where government was giving cash subsidies to beneficiaries directly—as in the case of scholarships or wages under the employment guarantee scheme or pension, the implementation of Aadhaar was quite straightforward: just credit the bank accounts connected with Aadhaar numbers of beneficiaries.

What Aadhaar essentially did was to unbundle identity, which meant that it could be plugged into other programmes where identity is a concern.

Take Rajasthan's Bhamashah scheme, a household-level identity system. Conceptualized in 2008, well before Nilekani started his work at UIDAI, it was named after a Rajasthani hero, an aide of Rana Pratap of Mewar. When Rana Pratap was running out of resources during his war with the Mughals, Bhamashah donated all his wealth to him. The scheme had three specific objectives: to empower women, to boost financial inclusion, and make the government-benefit programme more efficient.

All were well needed in the state. If there is one thing that strikes any visitor as hard as the Rajasthani sun, it must be the contrast between the opulent palaces and forts that dot several cities and the signs of poverty spread almost everywhere. Its economy thrives on extractive industries, providing most of the country's zinc, gypsum, marble; on tourism, leveraging on the achievements of the past; and on agriculture. It lags the Indian average in development indices, especially those related to women.

The Bhamashah programme aimed to provide a counter to these historic and societal trends. It designates a female member as the head of household (*nayak*) if the household consists of a female member over 21 years of age; otherwise, the oldest member is designated as nayak. Household bank accounts are in women's names too. Using Aadhaar and linked bank accounts, the cash benefits are directly transferred to the beneficiaries.

The roll-out of the programme was not perfect. There were design as well as implementation issues. When launched, the scheme was not mandatory, but made so in due course, depriving benefits to those who didn't have the card at the time of transition. The cash transfer limits imposed by Bhamashah were in one case lower than that of the scheme, and since all benefits were consolidated under Bhamashah, it essentially stopped some from getting their full entitlements. Besides, the limitations of biometrics caused problems for those with bad fingerprints.

Still, the overall impact was positive, according to a survey by MicroSave,[27] a financial inclusion consultancy, and Centre for Global Development, a Washington-based think tank. The report said the reforms have spurred financial inclusion (nearly all households now have at least one bank account; most have two or more). While there are some indications that women are getting a better deal, there is still a long way to go, in that mobile phones and finance are often still controlled by the male members of the family, even though on paper women are designated as the head. The Rajasthan government plans to study the social impact of the project in the near future. The perceived quality of government services had improved too, particularly in three schemes: PAHAL (consumer cooking gas cylinders), pensions, and the PDS.

The survey also found that in some cases people were not aware of some of the government programmes. For example, only 39% were aware of the state's insurance scheme, the Bhamashah Swasthya Bima Yojana, even though those who have availed it have expressed their satisfaction.

Amit Shukla, an engineer who has worked in Accenture and KPMG, says that 'a lot of problems with the government programmes arise not because the policy is bad, but because the implementation is bad. Often, citizens are not even aware of the schemes that could benefit them. Making it easy and actionable could solve much of the problem.' In 2016, Shukla co-founded EasyGov, an initiative that connects citizens with governments, and allows easy access to its programmes.

'Identity is just one part of the solution. To solve the problems, samaj, sarkar, and bazaar—the society, state, and the markets—should come together,' says Shankar Maruwada.

In India, much dust has been kicked up in the debates on the exact amount of cost savings thanks to Aadhaar. Politicians and bureaucrats, while trying to make a case for Aadhaar, often quote a figure from World Bank that said the project has saved the government $11 billion. However, activists questioned those numbers. In a column, Jean Drèze and Reetika Khera, argued that the amount that the original report referred to was not savings, but the total cash transfers, and that the subsequent justifications for arriving at a similar number were riddled with errors, such as misattributing some of the savings to Aadhaar, even where Aadhaar played no role.[28]

As we saw from the earlier examples, given that Aadhaar is only one part of a vast system, any attempts to find out the exact savings from Aadhaar would demand that we make too many assumptions. For acceptable quantitative evidence on the impact of broader systems change, one might look at randomized trial studies, such as the one conducted by economist Sendhil Mullainathan.[29] We couldn't access any such study on Aadhaar as of writing this. Most of the widely circulated surveys are often by the vocal critics of Aadhaar, and get dismissed by its supporters as being too biased.

But what has been clear from the ground so far is that to redesign a public goods distribution system, especially shifting some of the subsidies to direct benefit transfer, demands a reliable identification system. In fact, it would need to start with one.

Consider the Give It Up campaign, an initiative launched by the Modi government asking people who don't need cooking gas subsidy to give it up. The campaign was launched in 2014. According to oil minister Dharmendra Pradhan, by April 2016, over 11.3 million people, or 6.8% of the total 166 million customer base, had surrendered their subsidies and were paying market rates.[30] Contrast this to the deduplication of connections against Aadhaar database to check if there were ghosts. The exercise culled out about 680,000 duplicate or fake connections, a fraction of what was achieved by simply requesting citizens.

Yet, requesting citizens to give up subsidies could not have happened without first shifting to DBT, which was one of the outcomes of the rollout of Aadhaar. In 2009–10, India was consuming 14,000 metric tonnes of cooking gas every year, with close to a third as imports.[31] There was no restriction on the number of cylinders a household could consume. Oil marketing companies typically bought LPG at market rates, bottled it in cylinders and sold the cylinders to distributors at a subsidized cost that was determined by the Ministry of Petroleum and Natural Gas. The distributor took a commission and delivered it to the residents, who paid the subsidized cost plus delivery fee. The oil marketing companies got the subsidy amount from the government.

The system was leaky, thanks to the difference between the subsidized rates and the market price. While subsidized cooking gas cylinders were meant for households for domestic consumption, a good number of them were sold, mostly to commercial establishments, at the market prices by distributors. This also created a general shortfall of cylinders for households.

The government piloted a new way of giving LPG subsidies in Mysore. It capped the number of cylinders to six a year. And instead of paying the subsidized rate, customers would pay the market rate and get its benefits directly from the government into their bank accounts. There was resistance to the cap of 6 cylinders, and it was eventually increased to 11.

By most accounts, the Mysore pilot was a success, and it was scaled up to 291 districts. By January 2014, nearly 100 million LPG customers were getting their subsidies directly in their bank accounts.

This initiative too faced resistance. In Kerala, LPG distributors lobbied with their politicians to convince the Centre that people were not happy with the change, and that it might cost them the elections. Besides, there were concerns that the scheme was being rolled out too fast, including in the districts that had low Aadhaar enrolment rates. The government then put an end to it.

After a new government came to power, it not only revived the scheme, but also stepped on the accelerator, launched its Give It Up campaign, and effectively turned it into a publicity campaign for itself.

The Core of Aadhaar

'This is not a glamorous disease,' Dr Zarir F. Udwadia told his audience in a TedX speech in December 2016. 'It kills you slowly and agonizingly, ravaging your lungs. It stops you breathing, makes you cough up blood. Look, your friends and family will shun you because it's infectious. It doesn't require the mosquito vector of malaria or dengue. It doesn't need the sexual intimacy of HIV/AIDS. This is the perfect assassin. All it takes is a cough. Millions of infectious droplets are released and all you folk in the front row could be infected.'

Dr Udwadia, a Mumbai-based pulmonologist, was talking about tuberculosis (TB), on which his research is well known. He reminded the audience that deaths from TB alone far outnumber those from smallpox, plague, cholera, malaria, AIDS, and influenza—all put together. In India, he said, TB exists on an epic scale. 'It's our biggest public health problem. It refuses to go away. India houses the most TB patients in the world. TB kills the most Indians globally. One Indian dies of tuberculosis every minute.' India also produces 150,000 patients with drug-resistant TB.

The World Health Organization says:[32]

> Multidrug-resistant TB (MDR-TB) remains a public health crisis and a health security threat. WHO estimates that there were 600 000 new cases with resistance to rifampicin—the most effective first-line drug, of which 490 000 had MDR-TB. Globally, TB incidence is falling at about 2% per year. This needs to accelerate to a 4–5% annual decline to reach the 2020 milestones of the End TB Strategy.

One of the most effective ways to fight MDR-TB is with a methodology called DOTS, or directly observed treatment, short-course. One of the key elements of DOTS is a standardized reporting and recording system. This is needed to keep track of those who don't complete the full course, and thus turn the bacteria resistant to the treatment.

One organization in India, Operation Asha, co-founded by a doctor and former IRS officer, developed an e-compliance system that requires the community health workers who administer the medicine and the patients to authenticate using fingerprints or iris. They built a laptop-based system first with Microsoft Research, and later upgraded it for tablets. When a dose is missed, an alert is sent to the patient, the community worker, as well as a supervisor. This could bring down the non-compliance rate to less than 3%. A Columbia University team that studied the programme concluded that a patient using this system is 3.17 times more likely to be cured of tuberculosis compared to other patients.

Biometric identification and authentication are often used for security reasons. Before Aadhaar adopted biometrics, the biggest use case was the US-VISIT programme, which essentially used it as a security measure. The area that most people tend to associate biometrics with is forensics, which again has security connotations. The way Operation Asha used biometrics is fundamentally different from the way many tend to think of the technology. Neither the patients nor the community workers were in any way conspiring to skip doses. The goals of those who gave biometrics and that of the system were aligned.

It was in this spirit that Nilekani and his team chose biometrics to be the core of their identification platform.

First of all, they knew that it had to be inclusive. Which meant that it had to include millions of people who did not have a document that they could rely solely on. There would be many who did not have any proof of their existence so far, many who did not have a birth certificate, because their births were never registered, and they did not even know their birthdays; many who did not have a proof of address because they were homeless, and lived under bridges or behind temples or by the side of a mosque or a church.

Formal identification—outside one's social circles—has always depended on what someone possessed or what someone knew. For millions of Indians, it was not an option beyond their villages, districts, or states. One of the goals of UIDAI was to give them an ID that would be valid across the country, a way to authenticate themselves, irrespective of whether they are in the North-east, Kashmir, or Kanyakumari.

Besides, it was to be a foundational ID, unlike say a driver's licence or a passport, handed out only after determining eligibility. In the case of Aadhaar, as mentioned earlier, it did a single job.

Biometrics would also allow the system to be optimally ignorant about a person—without having to collect too many demographic and other details. Non-biometric identification methods often have to depend also on digital trails, for example. That's a reason why Google or Apple sends you a mail or an SMS when you login from a new IP address. That's also a reason why your credit card company might call you to confirm if you made a specific transaction, if it happens to be an outlier. Biometrics, in most cases, has to just depend on biometrics.

The primary reason for this dependence on biometrics, however, was the fact that India does not have a reliable birth or death registry. According to one statistic only 80% births are registered.[33] (In a country of a 1.2 billion people, with a large number of internal immigration, illiteracy, and poverty, it would be impossible for anyone to honestly claim that all births are registered.) While India is putting in place a system that is getting better at birth and death registries, UIDAI could not have created a unique identity platform without biometrics—given the urgency. (As Nilekani explained to an audience in Bengaluru, 'You cannot tell the poor, wait for 30 years and we will give a good identification system.')

However, biometrics comes with its own limitations. Not everyone has biometrics—such as those who lost their fingers or fingerprints or iris to accidents or disease or old age. Biometrics is also probabilistic, which means there could be false negatives and false positives. Besides, users have to pass a number of hurdles in an online authentication system—the devices have to work, the biometric capture should be good enough, the telecom network needs to be strong—compared to paper-based or even smartcard-based systems.

Technology can be a great tool for inclusion. But technology can also exclude people. Uber can help you find a cab fast, but that presumes that you own a smartphone, that you have access to it, and it's in working condition and has Internet connectivity. Uber excludes those who don't have access to a smartphone. The same is true of WhatsApp. WhatsApp can help you connect with your peer group and family members, it can ease the flow of information (and misinformation), but it again excludes people who don't have smartphones.

Such an exclusion triggered by technology can spill over to other domains. Once you are excluded from Uber—assuming that Uber is the best option to go from point A to point B, that is—you are forced to take the second best option. And you bear the cost in terms of time, effort, or money. Similarly, being excluded from a WhatsApp group keeps you out of your social group and the benefits that you can get from being a part of the group.

This is true of Aadhaar also. Some exclusion is inherent in the design of Aadhaar because of the use of biometrics. In fact, this was the first problem they had to consider once they decided to use the technology. Some of the initial trials were done in places that had a high concentration of manual labourers (which increased the chance of erasing fingerprints) and in leprosy homes.

As a result, UIDAI introduced biometric exceptions for enrolments. This had unintended consequences. Knowing that giving exceptions can improve the speed of enrolment, some agents began using this feature even in cases where people had good biometrics. (Biometric exceptions also allowed one agent to get an Aadhaar number for his dog, Tommy Singh, and another to get one for Lord Hanuman. Both happened in the early days of Aadhaar, and the checks and balances included eventually made it more difficult for such practices.) According to a reply in the parliament, over 50,000 enrolment operators have been banned for not following the process until now.

The problem with any kind of gatekeeping is that if you keep the door open too wide, you let a Tommy Singh in; but if you hold it too tightly you keep genuine cases out. Ultimately it's not just about how the system is designed, it's also about how well trained the last-mile operators are—not merely to get enrolment done in a fast and efficient way, but also to do so in an inclusive and humane manner.

The result of such difficulties was felt most by the poor and the disabled. For example, M. Dayalan, a visually impaired person in Tiruvallur district, was unable to get his disability pension because he couldn't get an Aadhaar number despite trying to enrol twice. At Leprosy Hospital in Bengaluru, six patients couldn't enrol for Aadhaar because the disease had eaten away their fingers and eyes—and as a result stopped getting their pension. It took several attempts by the doctor-in-charge of the hospital before they could get their numbers by exception.

Many of the problems related to biometrics go back to the point about Aadhaar being a Lego block. We saw how it can work, and enable the system to be more efficient. Given the limitations of biometrics, sometimes these Lego blocks can end up failing to plug into the system, if it's solely dependent on biometrics for authentication. This was why Babu, who was aware of these limitations, had incorporated exception handling in the form of mobile OTP and manual interventions in Krishna district. That helped.

Biometric technology is improving. Thanks to a wide variety of applications—in part driven by Aadhaar itself—biometric scanners have dropped in prices. Fingerprint and iris scanners are getting embedded in smartphones. Some of these scanners include a liveness filter, which can detect if a fingerprint is cloned and thus can improve security.

UIDAI has been making improvements as well. It launched a biometric lock to reduce the risk of misuse. It announced plans to launch facial authentication, which can be used in conjunction with fingerprints, to improve the chances for those who have bad fingerprints. A combination of such improvements, natural evolution of biometric devices, and better design of the systems that use Aadhaar promise to bring down some of the negatives associated with the use of biometrics.

However, it's worth remembering that even if the technology improves exponentially, the 'Lego block' logic will hold good. Aadhaar will be a single block in a technology solution, which in turn will be a part of a solution set—legal, economic, managerial—hoping to solve the problem.

It Depends on Inbuilt Incentives

When Aadhaar was launched, Nilekani and his team were working under the assumption that it would be voluntary; they knew that to attract people, they have to see value in it. But then, by itself Aadhaar had no value. As Pramod Varma explained to his audience in the UN, it does nothing by itself. They would see the value only when it was used in solutions.

The core team that designed and built Aadhaar had an inherent trust in the mechanics of the market (while being aware of market failures and the role of a visible hand to correct some of those, a reason why they either joined the government or came together to work on the project), so they used incentives cleverly.

The technical architecture—the so-called hourglass (the Aadhaar number), bringing together several layers of applications and several devices above and below it—reduced the scope of what Aadhaar by itself would do.

By narrowing the scope of what they would do, the team could think broadly about the role others would play. Srikanth Nadhamuni, head of technology for UIDAI, says that when he was studying biometric systems across the world, he found that many got locked into one provider, and once that happened the dependence on the provider crept in. In effect, the customer can be held to ransom because there is only one vendor. UIDAI didn't want to get into that mess. It gave the specifications, built the application program interface (APIs) to make sure different systems spoke to each other, and let the vendors compete.

In the broader market such competition typically leads to lower prices and better quality, and so it was with UIDAI. Prices came down. This had an impact not just on the cost of the project—unlike big government projects, there were no cost or time overruns—but the prices came down globally as well.

Consider yet another case. At the core of Aadhaar's biometric system is a deduplication algorithm that figures out whether each biometric image is unique. It's huge, both in terms of scale and complexity. To give the billionth Aadhaar number it has to check a person's biometric data for uniqueness against 11.99 billion biometric images (fingerprints and irises). No system is so perfect that it will generate zero error rates. UIDAI enrolled three different vendors to do the task, each using their proprietary algorithm. Those who did better were dynamically allocated more work and thus more money, creating competitive pressure.

These principles helped Aadhaar achieve scale without cost or time overruns. However, the value of Aadhaar is not in people having a unique number, but in using it. For Uber or Ola, the number of app downloads is important, but not sufficient. The value really is in the number of times customers—the drivers as well as passengers—use it. To get them to use it, these companies constantly tweak their incentives until it becomes a habit, and the full power of networks kick in. In the long run, however, their success depends on the value it consistently delivers to the stakeholders. For instance, and at a more fundamental level, the value of GPS depends on the number of applications that use it. In other words, the number of businesses such as Google and Uber.

In the government, though, the push for its use has to come from the drive of bureaucrats such as A. Babu and that of politicians such as the ones who launched Aadhaar as an inclusion tool. If revenues and profits provide the feedback to the private sector; for the government, it's the feedback from civil society that reveres principles such as transparency and inclusion.

One of the best examples of how this dynamic works comes from Uganda. In the 1990s, Uganda was the focus of attention for development economists across the world, thanks to an innovative experiment.

It started with two researchers: Ritva Reinikka from the World Bank, and Jakob Svensson from Stockholm University. They were trying to find out if the grants that Uganda's central government gave to primary schools actually reached their destination. Just as the Indian government gives funds to schools to provide midday meals based on the number of students a school has, the Ugandan government allocates a capitation grant per enrolled student to schools to buy textbooks and instructional material, to maintain the buildings, and so on. The money is transferred to the district administration and from there to the schools. Or that's how it was supposed to work.

However, when Reinikka and Svensson surveyed 250 schools about the inflow of money and checked it against the outflow from government coffers, the results were stunning. Between 1991 and 1995 only 13% of the funds meant for the schools reached them. The rest was diverted for other purposes, or it simply disappeared.

The results of the survey spurred the Ugandan central government into action. It could have come up with a complex solution to fix the problem, involving probes and investigative officers. Instead, it opted for a simple, even counterintuitive, way. It started publishing monthly transfers of public funds in newspapers and splashed them on radio. And it asked the schools to post notices on the inflows of funds. The results were dramatic. 'Instead of about 20% in 1995 over 90% of the intended capitation grants reached the schools in 1999,' the researchers wrote in a 2001 paper titled 'Explaining Leakage of Public Funds'.[34]

What exactly was going on?

The problem with much of public spending, especially in countries that don't have strong institutions, is that the government is often in the dark about where the money goes—the scale of operations is that huge. At the time of the survey, the government was giving grants to about 8,500 primary schools. They were funnelled through the district administration, which posed what's called the 'agency problem'—the officials didn't work with the best interests of the schools in mind. It led to leakages. While everyone knew there were leakages, until this survey was done, it wasn't even quantified. The government didn't know, and neither did the schools. There was no transparency.

While the grants were supposed to be based on the number of students a school had, in reality it didn't work that way. The schools that had more involved parents—ones who actively participated in the Parents Teachers Association (and they were typically rich) received most of the grants they were meant to get. Similarly, the schools that were showing better grades received their grants, because the top officials tend to visit better performing schools. But mostly, it depended on the relationship individual school principals had with the district administration officials. The better the relationship, the lesser the leakages. The other schools were neglected. The money they were supposed to receive went elsewhere.

When the government started publishing the outflow of funds in newspapers and asked schools to publish inflows, suddenly there was a new level of transparency. The bargaining power of almost all school principals went up, because they had data with them. In their book *Poor Economics*, economists Abhijit Banerjee and Esther Duflo described what happened thus:[35]

> About half of the headmasters of schools that had received less than they were supposed to had initiated a formal complaint, and eventually most of them received their money. There were no reports of reprisals against them, or against the newspapers that had run the story. It seems that the district officials had been happy to embezzle the money when no one was watching but stopped when that became more difficult.

The Push by the Government

In October 2017, at a panel discussion in Bengaluru, Charles Assisi, one of the authors of the book, quizzed Congress leader Jairam Ramesh on why he and his party changed its stance on Aadhaar. Clearly, the mood on Aadhaar had changed significantly since 2017. In the toilets of the hotel that the panel discussion took place, a group of activists had dropped Post-it notes that said 'Aadhaar is slavery' into the urinals. That has been one form of protest against Aadhaar, and one could find it in every event in Bengaluru where Aadhaar gets discussed.

Ramesh's reply was elaborate, and it reflected what many have said about Aadhaar. He said he was excited about the project initially because it promised to fix the leakages in the system, and make government projects more efficient. He had implemented Aadhaar in his own department for delivering pensions. However, the character of Aadhaar had changed after the BJP took over. 'One part of Aadhaar is: how do you make government programmes more efficient. The other part of Aadhaar is: do you need Aadhaar for mobile phones? Do you need Aadhaar for bank accounts? Do you need Aadhaar for flying? I mean do you need Aadhaar for being born? Then there was a guideline that was issued that you need Aadhaar for a death certificate!' As he said this the audience broke into an applause.

When Jairam Ramesh was making those arguments he was not merely playing to the gallery. His bewilderment at the widening scope of Aadhaar was probably genuine. People's attitude towards specific issues are often shaped by their own lives and the histories of institutions they are associated with. As we saw earlier, the Congress party wanted a national identity programme much before Nilekani joined the team. However, their ambition was limited to establishing citizenship.

Citizenship is a right and a matter of pride, Congress party said in its election manifesto of 2009. 'With the huge IT expertise available in our country, it is possible to provide every Indian with a unique identity card after the publication of the national population register in the year 2011.'

This partly explains why Nilekani had to fight some of the battles even within Congress. The buy-in for the project was not evident across the leadership. It had a few believers—primary among them Pranab Mukherjee. Manmohan Singh was mostly non-committal. P. Chidambaram (at least at one point) was against it. Many looked at it with suspicion because they thought it was encroaching on the Home ministry's turf, which had limited use cases for its identity cards.

Contrary to a popular perception that Modi was against Aadhaar—which is broadly based on a well-publicized tweet and the election rhetoric—he was for it, and it would have been evident if people looked at what he was doing in Gujarat, rather than what he was saying. Another equally popular perception about Modi is that he changed his mind about Aadhaar, and like a recent convert, pushed the pedal on it even faster than the Congress. But a far-reaching national identity cards programme was in the BJP's agenda long before Modi was considered a prime ministerial candidate. Its 2009 election manifesto—which had photographs of A.B. Vajpayee, L.K. Advani, and Rajnath Singh on the cover—went far more in detail into the subject of a national identity card:

> The BJP will launch an innovative programme to establish a countrywide system of multipurpose national identity cards so

> as to ensure national security, correct welfare delivery, accurate tax collection, financial inclusion and voter registration. Voter identity cards, PAN cards, passports, ration cards and BPL cards are already in use though not all with photo identity. The NDA proposes to make it incumbent for every Indian to have a National Identity Card. The programme will be completed in three years.
>
> The National Identity Card will contain enough memory and processing capabilities to run multiple applications. Through it the NDA will ensure efficient welfare delivery and tax collection. The card will also be linked to a bank account. All welfare payments, including widow and old age pensions, through the wide range of schemes such as Mother and Child support/Kisan Credit, Students Assistance and Micro-Credit will be channelised through the National Identity Card. The card will make it possible for individuals to save and borrow money; for farmers to get bank credit, also establish accurate land titles data.
>
> The National Identity Card will also strengthen national security by ensuring accurate citizen identity, thus tracking illegal immigration. All financial transactions, purchase of property and access to public services will be possible only on the basis of the National Identity Card which will be made forgery and hacking resistant.

After he became prime minister, Modi was simply fulfilling the promise his party made to voters in 2009. Only, Aadhaar was not a card. It came in a much more flexible form.

The BJP's desire to connect Aadhaar with other services beyond subsidies are not without intellectual backing. During a visit to India a year after serving as RBI governor, Raghuram Rajan—he had gone back to teaching at the Chicago Booth School of Business—was his usual candid self while taking questions on economic policies, including demonetization, which was ostensibly carried out to tackle black money. Rajan said there were better alternatives. Linking Aadhaar to financial investments and bank accounts gave tax authorities better visibility of bad actors (and there are technology solutions to bring down its misuse). 'By moving to an Aadhaar-based system, you can get far better compliance on the flows,' he said.[36]

Again, in the policy circles, Aadhaar was seen as a much-needed tool for financial inclusion. A board member at RBI told us that once there were criticisms that the KYC requirements were too stringent, that it kept people out of the banking system. However, it turned out that there was need for that in the age of digital banking, and Aadhaar made the entire process easy and convenient.

However, the difference between Congress and BJP was not just about scope but also about speed. A six-panel cartoon by Sandeep Adhwaryu[37] captured the difference between Manmohan Singh and Narendra Modi. In two of the three panels devoted to him, Singh is seen reading books on surgery in an emergency room as a patient waits in panicky distress. By the time he starts his work, by zeroing on where to cut, the patient is dead and gone. Modi, in contrast, is all set to cut the patient with a saw in the first panel, rummages through his innards in the second, and starts reading ABC of Surgery in the third, standing next to the dead patient.

Take a worldview that believes in the linking of Aadhaar to many services, and on top of it, add the obsession with rashness. What we are left with is a pressure to link Aadhaar with multiple services.

In some ways, this struck right at the roots of some of the inherent assumptions made by the team that designed and rolled out Aadhaar. For one, Aadhaar was built on the assumption that it would be voluntary, except when you are accessing subsidies from the government. The way to get out of Aadhaar was to simply say, 'I don't want any subsidy from the government.' Nilekani told us as much in 2013.

It was not just Nilekani; many in the core team felt the same way. They were confident that people would start using it because Aadhaar was so much more convenient and safer than the other options. One of the reasons Nilekani reached out to Shankar Maruwada early on was because he knew that being voluntary, the idea had to be marketed. The team went around to different government departments knowing that they had to persuade others by reason, and not the power of their positions. Many were volunteers themselves, and had no power to speak of.

A bureaucrat who worked closely with him during his UIDAI days told us that Nilekani felt 'somewhat betrayed' when the nature of Aadhaar turned from voluntary to mandatory. 'Throughout his term as UIDAI chairman, he kept telling everyone it was voluntary, and suddenly, by forcing people to link Aadhaar to every service, it was mandatory.'

However, when questioned by the media in public, Nilekani often took refuge in legalese instead of revealing his personal opinions. 'If the government wants to make it mandatory for PAN to weed out fake cards, what's the problem there?' he once asked without elaborating.

In 2018, when journalist and television anchor Vir Sanghvi asked Nilekani if he would be happy if the Supreme Court (which had just heard a case questioning the constitutionality of Aadhaar) decided that Aadhaar should not be mandatory, he replied in affirmative. Sanghvi seemed surprised, as Nilekani went on to explain, 'My view is that the real use of Aadhaar is beyond all this, when it is used for delivering online benefits, for innovations on the top of this. So, if it's not mandatory for something, I am fine with it.'[38]

Using technology is often like shopping for a pair of shoes. You don't pick shoes that are big for your feet because you had a bad experience with a small pair of shoes. You go for the right fit. By many accounts, the government ignored this, and made a similar mistake. It went on an overdrive.

An IAS officer told us that in his experience this was probably the most well-intentioned government. However, there were too many directions from the Centre, taking no cognisance of what was happening on ground: No one is pausing to ask why. Everyone is rushing to find out how, and how fast something can be done.

He shared a joke that was floating around in bureaucratic circles.

> Senior officer: We need to do something innovative. Do you have any suggestions?
>
> Junior officer: Yes, sir. We can ask people to link their Aadhaar numbers.
>
> Senior officer: And then what?
>
> Junior officer: Well, we can think about all that later.

Notes

1. Shefalee Vasudev, 'Nameless Death', *India Today*, 8 March 2004, http://archives.digitaltoday.in/indiatoday/20040308/controversy.html, viewed on 16 July 2018.

2. Zonal Integrated Police Network, http://zipnet.in/index.php?page=un_identified_dead_bodies_search&criteria=browse_all, viewed on 16 July 2018.
3. Akshaya Trust, http://www.akshayatrust.org/, viewed on 4 July 2018.
4. Gurcharan Das, *India Grows at Night* (New Delhi: Penguin Books, 2012).
5. Farzana Afridi, 'The Impact of School Meals on School Participation: Evidence from Rural India', Indian Statistical Institute, 2010, https://www.isid.ac.in/~pu/dispapers/dp10-02.pdf, viewed on 4 July 2018.
6. T. Jacob John and Vipin M. Vashishtha, 'Eradicating Poliomyelitis: India's Journey from Hyperendemic to Polio-free Status', *The Indian Journal of Medical Research* 137, no. 5 (2013).
7. Atul Gawande, 'The Mop-Up', *New Yorker*, 12 January 2004.
8. PTI, 'Rajiv Gandhi's Popular 15 Paise Remark Finds Mention in Supreme Court Verdict', *Indian Express*, 9 June 2017, http://indianexpress.com/article/india/rajiv-gandhis-popular-15paise-remark-finds-mention-in-sc-verdict-4696740/, viewed on 26 June 2018.
9. Planning Commission, 'Performance Evaluation of Targeted Public Distribution System, 2005', http://planningcommission.nic.in/reports/peoreport/peo/peo_tpds.pdf, viewed on 26 June 2018.
10. Rediff.com, 'PM Sets Up Monitoring Group to Track Big Projects', 13 June 2013, http://www.rediff.com/business/report/pm-sets-up-monitoring-group-to-track-big-projects/20130613.htm, viewed on 26 June 2018.
11. Raghuram Rajan, 'Finance and Opportunity in India', RBI, 11 August 2014, https://www.rbi.org.in/scripts/BS_SpeechesView.aspx?Id=908, viewed on 26 June 2018.
12. Sangeet Paul Choudary, *Platform Scale* (Platform Thinking Labs, 2015).

    Geoffrey G. Parker, Marshall W. Van Alstyne, and Sangeet Paul Choudary, *Platform Revolution* (New York: W.W. Norton, 2013).

13. John Mauldin, 'India's Tech Revolution Has Already Left the West Behind—It's the Best Investment Opportunity Now', *Forbes*, 9 April 2017, https://www.forbes.com/sites/johnmauldin/2017/04/09/indias-tech-revolution-has-already-left-the-west-behind-its-the-best-investment-opportunity-now/#3dcf7c2b2360, viewed on 26 July 2018.
14. PTI, 'Aadhaar Critical Step in Enabling Fairer Access: UN', *Economic Times*, 1 December 2016, www.economictimes.indiatimes.com/articleshow/55723039.cms, viewed on 26 June 2018.
15. Lauren Medley, 'How India's Unique ID System Is Changing Lives', World Bank, 3 April 2013, https://blogs.worldbank.org/voices/how-indias-unique-id-system-changing-lives, viewed on 26 June 2018.
16. Jeanette Rodrigues, 'India ID Program Wins World Bank Praise Despite "Big Brother" Fears', *Bloomberg*, 16 March 2017, https://www.bloomberg.com/news/articles/2017-03-15/india-id-program-wins-world-bank-praise-amid-big-brother-fears, viewed on 26 June 2018.
17. Shofar Blast.org, 'Digital ID Network unveiled at ID2020 Summit in NY', 26 July 2017, https://shofarblastblog.com/2017/07/26/digital-id-network-unveiled-at-id2020-summit-in-ny/, viewed on 18 August 2018.
18. Pramod Varma, 'Identity as a Platform for Change', UN Headquarters, 2017, http://webtv.un.org/watch/2017-id2020-platform-for-change-summit/5476783692001, viewed on 15 July 2018.
19. Varma, 'Identity as a Platform for Change'.
20. Varma, 'Identity as a Platform for Change'.
21. Richard Feynman, 'Cargo Cult Science', Caltech, 1974, http://calteches.library.caltech.edu/51/2/CargoCult.htm, viewed on 5 July 2018.
22. Robert Pirsig, *Zen and the Art of Motorcycle Maintenance* (New York: William Morrow, 1974).
23. Kentaro Toyama, *Geek Heresy: Rescuing Social Change from the Cult of Technology* (New York: Perseus Books Group, 2015).

24. Kentaro Toyama, 'Technology Is Not the Answer', *The Atlantic*, 28 March 2011, https://www.theatlantic.com/technology/archive/2011/03/technology-is-not-the-answer/73065/, viewed on 5 July 2018.
25. Planning Commission, 'A Hundred Small Steps', 2009, planningcommission.nic.in/reports/genrep/rep_fr/cfsr_all.pdf, viewed on 5 July 2018.
26. Nandan Nilekani, *Imagining India* (New Delhi: Penguin Books India, 2008).
27. MicroSave and Centre for Global Development, 'Impact of Bhamashah on Digital Governance Reforms in Rajasthan', 2017, http://www.microsave.net/files/pdf/171212_Household_Perception_Impact_of_Bhamashah_Digital_Governance_Reforms_in_Rajasthan.pdf, viewed on 5 July 2018.
28. Jean Drèze and Reetika Khera, 'Aadhaar's $11-Billion Question', *Economic Times*, 7 February 2018, https://blogs.economictimes.indiatimes.com/et-commentary/aadhaars-11-bn-question/, viewed on 5 July 2018.
29. Dean Karlan, Sendhil Mullainathan, and Benjamin Roth, 'Debt Traps for Micro-Entrepreneurs in Chennai, India', J-PAL, January 2018, https://www.povertyactionlab.org/evaluation/debt-traps-micro-entrepreneurs-chennai-india-0, viewed on 20 July 2018.

    Anandi Mani, Sendhil Mullainathan, Paul Niehaus, and Anuj Shah, 'Evaluating Alternative Cash Transfer Designs in Kenya Using Behavioral Economics', J-PAL, https://www.povertyactionlab.org/evaluation/evaluating-alternative-cash-transfer-designs-kenya-using-behavioral-economics, viewed on 20 July 2018.
30. Gireesh Chandra Prasad, '11 Million Consumers Give Up LPG Subsidy', *Mint*, 22 April 2016, https://www.livemint.com/Politics/hTg1ZfFqNGkF9TgKs9Sa5K/About-a-crore-consumers-give-up-LPG-subsidy.html, viewed on 5 July 2018.
31. Ministry of Finance, Government of India, 'Interim Report of the Task Force on Direct Transfer of Subsidies on Kerosene, LPG

and Fertiliser', June 2009, https://www.finmin.nic.in/sites/default/files/Interim_report_Task_Force_DTS.pdf, viewed on 5 July 2018.

32. World Health Organization, 'Tuberculosis Fact Sheet', http://www.who.int/news-room/fact-sheets/detail/tuberculosis, viewed on 5 July 2018.
33. Arti Dhar, 'Births and Deaths Registration Still Low in India', *The Hindu*, 18 April 2013, https://www.thehindu.com/news/national/births-and-deaths-registration-still-low-in-india/article4630425.ece, viewed on 30 July 2018.
34. Ritva Reinikka and Jakob Svensson, 'Explaining Leakage of Public Funds', World Bank, 2001, http://www1.worldbank.org/publicsector/decentralization/Feb2004Course/Background%20materials/Reinikka2.pdf, viewed on 5 July 2018.
35. Abhijit Banerjee and Esther Duflo, *Poor Economics* (New Delhi: Random House India, 2011).
36. Sidhartha and Surojit Gupta, 'I Left Because There Was No Offer on the Table from the Govt: Raghuram Rajan', *Times of India*, 3 September 2017, http://timesofindia.indiatimes.com/articleshow/60341824.cms, viewed on 5 July 2018.
37. Sandeep Adhwaryu, *Times of India*, https://imgur.com/r/india/2aarwc5.
38. CNN-News18, 'Virtuosity: Nandan Nilekani and Rahul Matthan on Aadhaar, Privacy Laws', YouTube, 23 July 2018, https://www.youtube.com/watch?v=eU7z5gWJEpI, viewed on 28 July 2018.

4

When a Butterfly Flaps Its Wings

THIS IS A STORY ABOUT HOW INDIA ACCIDENTALLY crafted digital infrastructure that the world can emulate. It wasn't intended to be this way, but India Stack—a set of functions and tools that allows for third-party development—now exists and has grown too big to fail. Put it down to the Butterfly Effect,[1] if you will. The idea, postulated by American mathematician and meteorologist Edward Lorenz, is that if a butterfly flaps its wings in Rio de Janeiro, a tornado may occur in distant Florida many weeks later. We could well use this metaphor to understand India Stack.

It started life as Aadhaar, a project to provide every Indian resident with a unique identity number. But even as it was being deployed, it took shape as India Stack. In its current avatar, the Stack has five distinct elements.

- eKYC, or the electronic equivalent of a document that satisfies any entity's KYC requirements as mandated by the regulatory authorities.

- Aadhaar Payments Bridge, which turns an Aadhaar number into the person's financial address and resembles an email address.
- eSign, a digital signature that can be affixed on documents.
- Unified Payments Interface (UPI), a platform on which payment apps can be built and money transferred in much the same way an email or text message is sent through a mobile phone.
- A consent layer to share personal data (mark sheets, health records, financial transactions) with a bank, insurer, employer, or university for a limited time for a specific purpose.

When all of it is put together, people can authenticate their identity, sign documents digitally, store them in digital lockers, and transfer funds from one account to another using digital wallets. The stated outcome of this suite is to create a paperless, cashless, digital economy where middlemen are eliminated.

The implications of all that is possible with India Stack is obvious to those embedded in technology, in policy-making circles, and those familiar with the term 'Application Programming Interfaces'.

For the uninitiated, APIs are a set of definitions, protocols, and tools for building software. If you imagine a car to be a software component, then a car API would be a set of instructions on how to start the engine, accelerate, apply brakes, etc. A person who wants to drive the car does not have to bother about the mechanics at work under the hood. Their focus is on its functions. Similarly, if you imagine the kitchen at a restaurant as a software component, a kitchen API would be the waiter. You place your order, and the waiter fetches what you need. You need not worry about the complex operations inside the kitchen.

You use the waiter to get what you want. APIs work in a similar fashion. In short, APIs make it easy for developers to use technologies in the way graphical user interface made it easy for people to use applications like Word or Excel.

When the infrastructure that is the Internet got into the public domain and captured public imagination, the pattern repeated. People wanted to be on it and started to create content to seed it. Soon enough, there was a need to create something to index the content across websites and search all of it. Many contenders emerged to build a search engine. Google is now the world's most popular search application. But Google didn't stop with that. It created an API that allowed other applications to access its search function.

With that in place, Google built another application called Gmail. And a Gmail API for that. When search is embedded into email, life becomes easier for a consumer—and Google becomes a significantly more valuable company. The more products it built, the more they 'talked' with each other, the more valuable Google grew.

Similarly, India Stack is a bunch of APIs that can be used by anybody to create new applications. They can even create a profit-making entity like Google if they can think up ways to use their APIs innovatively enough.

Aadhaar allows authentication of a person's identity through a federated model—via biometrics, OTP, PIN, etc. That is all Aadhaar can do. A start-up building an application that needs someone to be authenticated can access Aadhaar through an API.

Once authenticated, an individual may want to open a bank account. Somebody else may want to deploy it for an altogether different purpose—like provide documentation to get a telephone connection. The eKYC API does the job.

After having opened a bank account, somebody may want to acquire a loan to buy a truck. To do that, documents must be signed. The e-Sign API can be used to create a digital signature in such cases.

Or a bank may want to examine an individual's credit history and source of income before disbursing a loan. These documents can be fetched from a digital locker, built using the DigiLocker API.

If a loan is approved, money can be transferred instantly from an app that uses the Unified Payments Interface application. On UPI, it is possible to add more functions—like a digital wallet to carry digital money so the bank account does not have to be accessed all the time.

The Indian government's version of UPI is called the Bharat Interface for Money (BHIM) app. There are many payment apps around, including those built by Google and WhatsApp. Much money is being spent and bitter battles are being fought to dominate this space.

When India Stack starts talking to the Goods and Services Tax (GST) system, using GST Network (GSTN) APIs, the world's largest tax-based APIs, its value goes up manifold.

The Early Days

When the founding team of Aadhaar got together, their mandate was clear: Create a unique identity for over a billion people that all agencies in India can accept with confidence.

That is why when the first Aadhaar number was issued, the authentication API and an Aadhaar-enabled Payment System (AePS) was put in place as well at the National Payments

Corporation of India (NPCI) through which all banks could speak to each other (in technical terms, be 'interoperable'). This wasn't an afterthought—the stated intent of the system was to transfer benefits to those who could be identified. What nobody was aware of though was the scale at which it could be done.

That, Pramod Varma recalls with much amusement, is because 'at that time, there may have been about 100 users in the country and nobody really cared about it. You can't do anything with a technology until some critical mass is reached.'

The first visible sign of scale started to make itself apparent in 2011, when the Indian Bankers Association asked if the core banking system on which they worked could connect to AePS. But before that, there was much behind-the-scenes evangelizing to be done.

By way of example, banks that had built entire sales processes over the years to acquire customers were not keen on interoperability. Their representatives argued it had taken them much time and money to woo customers. If interoperability were allowed, new banks that hadn't put in as much grunt work could take their customers away without investing in new branches or full-fledged ATM outlets.

A.P. Hota, the managing director of NPCI until August 2017, recalls those days and the conversations that followed. NPCI was formed around the same time as UIDAI. One of its mandates was to figure out how technology and innovation can be deployed into all retail payment systems in the country. It had the support of the RBI and had earned the trust of bureaucrats from the Finance ministry as well.

When the RBI issued a diktat that interoperability be allowed, banks with a presence in rural India protested about the fees. They argued that their customers withdraw on average

Rs 200–300 each time. Imposing Rs 15 as a fee was too high a cost for them. Protracted discussions later, it was brought to the RBI's notice.

Eventually, between RBI and NPCI, they asked all the stakeholders in the ecosystem if they were amenable to accepting Rs 5 as a fee for transactions up to a prescribed amount for certain categories. After that, they could charge Rs 15. Everybody agreed.

Solving this was only the tip of the iceberg. The analysts at NPCI monitored 2 million transactions emerging out of rural India. Their enquiries revealed that the 4-digit PIN number for a debit card was inevitably 1234. They thought it a serious security risk and alerted the Finance ministry and the RBI. Officials at all three entities got into a huddle. They could see the immediate risks.

The government was keen to implement the Pradhan Mantri Jan Dhan Yojna (PMJDY),[2] a financial inclusion programme launched on 15 August 2014 by Prime Minister Modi. It was announced with much fanfare and everybody knew it would take off in a big way. By February 2018, it had 310 million accounts and $12 billion parked in it. The risk weak PIN numbers posed had to be mitigated. After much deliberation, they arrived at a consensus.

The RBI passed an order that from 1 January 2016 all new terminals set up by any bank ought to have biometric authentication capabilities. This order included micro-ATMs as well. (These are handheld versions of ATMs that look like a credit card machine. These devices are used to reach the hinterlands by an institution's banking correspondent network, so people can avail of basic banking services, including cash withdrawals.)

While biometric authentication is not mandatory at all ATMs, it is mandatory for all financial inclusion projects that involve taxpayers' money and public expenditure. It got implemented, Hota says, only because the biometrics had already been captured when Aadhaar was set into motion. In his reckoning, the success rate is 93% and improving.

When asked what about those who might get excluded, he said that they can be taken care of by back-up mechanisms. More importantly, though, he would much rather have a secure system that works 93 times out of 100 than have an unsecure system that can be deployed to siphon money away from the poor.

In any case, he says, technology is improving every day and it is only a matter of time before exclusions are fixed completely.

Our conversations in places across the country—from metros such as Delhi, Mumbai, and Benglauru, or the smaller cities such as Kochi in Ernakulam district or a small town like Kodungallur in Kerala, the most literate Indian state—suggest that technical failures are not the only reason why exclusions occur. Sometimes, people in the system actively collude to make it happen. Take shops established on account of the PDS that were originally intended to benefit the most needy in India, for instance. The government insists that all stores that issue subsidized food and other essential items must have biometric readers to authenticate an individual's identity. Some vested interests routinely tamper with these readers at a few outlets to ensure they fail. What it leaves at the other end are exasperated and needy people who curse a system that does not work.

A retired bureaucrat who did not want to be named said that this is the nature of the system. That this is a 'cat-and-mouse game' and problems of such kind cannot be exhaustively anticipated while building a system, but must be worked upon, and the system must be improvised even as it is being rolled out.

Know Your Customer, Digitally

With UID rolling out, the team posed itself another question: After identity is established, how may eKYC be done? This, because regulatory bodies insisted that to avail certain services, it is mandatory that businesses know the correct details about their customers. Aadhaar was architected such that it would only allow a person's UID number, name, and photograph to be shown—not their address or any other demographic details.

Not just that, Sanjay Jain, who was a part of the founding team at UIDAI, recalls that while they had thought up a digital identity that could be authenticated, many institutions, ironically enough, insisted on a photocopy of the document. The founding team had imagined an India where photocopies of people's documents are eliminated so they cannot be duplicated for illicit purposes. But neither the systems nor the regulations would permit it.

It fell upon Varma, Jain, and team to think through this. A potential solution, they imagined, is that an independent entity everybody in the ecosystem trusts certifies it digitally. If this is made possible, people can carry a digital eKYC certificate with them on their mobile phones and use it whenever needed.

But they concluded the idea was ahead of its time. People still had to wrap their heads around the idea of Aadhaar, a digital identity. A digital certificate would complicate things further.

So, they came up with what is being used now. If an individual offers consent, their details such as address, gender, and date of birth can be accessed by a third party. The technicalities of this were ironed out as early as 2012. But for an eKYC to be acceptable to a bank, a circular had to go out from the RBI. It wouldn't be until early 2014 that this was issued. SBI and Axis Bank were among the first to latch on to it. Early adopters could see the advantage when they did the math.

Arundhati Bhattacharya, the former chairman of SBI, puts things into perspective by pointing to how the PMJDY could take off at scale: Data pouring in showed 30% of all accounts at the bank now originated from the push created by the scheme. But 97% of these accounts had no money in them. This, she says, was partly because there was no way the bank could comply with the KYC requirements as mandated by the RBI. But after eKYC was allowed, zero-balance accounts dropped dramatically to 21%. On average, each account now has a four-figure amount in it, and when we last spoke with her, the bank was managing Rs 160 billion of deposits in these accounts.

'There is a latent demand [for bank accounts]. We could have done financial inclusion earlier. But we could not scale it. Aadhaar helped us scale. Without it, this would have been just another scheme. It allowed us to do the right things,' she says.

Jain has it that after factoring in all the costs, it could cost a regional bank up to Rs 1,000 to acquire a new customer. While there are various estimates doing the rounds for banks at differing levels of efficiencies, the data he has suggests that when operated at its optimum, this cost has come down to Rs 40 for a private sector bank. The more dramatic reduction, he says, is in the cost of acquiring customers for a mutual fund, from Rs 1,500 to Rs 50.

These are the kind of numbers that will get banking to grow by at least 7–8%, says Rajiv Anand, executive director, retail banking, Axis Bank. In his reckoning, when technologies like these are deployed by micro-finance institutions (MFIs), they can bring larger numbers of people into the system rapidly.

For an MFI to make a dent and operate at optimum levels, he says, it must acquire at least 100,000 people. What held them back until now was the cost involved in on-boarding people. With eKYC, this cost goes down. Much is possible now and new kinds of entities in the space can emerge, Anand says—ones that offer loans to women or small entrepreneurs in urban conglomerates across the country that operate in a specific industry, for instance.

On the ground, eKYC had its own issues. Unfamiliar with a new system, users sometimes did not know what they were consenting to. Airtel Payments Bank Limited (APBL) is a case in point. The first company in India to get a payments bank licence, APBL opened a large number of accounts and connected them to their customers' Aadhaar numbers, allegedly without their informed consent. Many Airtel mobile customers thought the company was doing Aadhaar-based reverification, but ended up agreeing to open a bank account. In response, UIDAI suspended APBL's eKYC licence and imposed a fine of Rs 25 million on Airtel.

Sign the Documents

Even as this solution was being deployed, Ram Sevak Sharma, who had taken over as the IT Secretary, had a question playing on his mind. If a UID and eKYC is possible, why not launch a digital signature as well? Or e-Sign as it is now called.

When Sharma looked around the digital landscape in 2014, what he saw didn't add up. Six hundred million people had been issued a UID; an IT Act was also in place. Yet, there were only 5 million digital certificates and 2 million digital signatures. This was possibly because those who had them were directors who sat on the boards of a number of companies and signed off at a number of places.

It was time, he thought, to consult Varma again. Varma was aware of the limitations of dongle-based digital signatures, and had given some thought to this as well. He could see the merit in Sharma's argument right away and got down to crafting a solution that could work in India.

Once he was done, Sharma and his team took it to the Law ministry. They were convinced this signature would adhere to all protocols as mandated by the IT Act, and a gazette notification was issued in June 2015[3] that an individual's digital signature can be accepted in India.

'What that means today,' says Varma, 'is that over one billion people can sign a legally binding document sitting anywhere in India without moving out of where they are. And you don't have to cut trees for paper.' A week after the notification went out allowing people to sign digitally, another note went out inviting private participation[4] from firms or individuals who may be interested in the business of issuing digital certificates.

'This threw the doors open to many providers to compete with each other so long as they adhered to standards prescribed by the Controller of Certifying Authorities [CCA]. More entities started to get into the ecosystem. The Securities and Exchange Board of India [SEBI, the securities market regulator] made it easier for people to trade in the stock markets, banks [found it easier] to open accounts and telecom service providers to sell SIM cards, now that a presence-less and paperless layer had become possible,' says Varma.

To understand the full import of what he means, it is pertinent that we go back to the metaphor we started with—that you can drive a car without knowing the complexities under the hood. Entities can access these functions without worrying about the complexities so long as they adhere to standards put in place by certifying authorities.

A digital leap was taking place. The government was inviting private entities to come in and help people create digital signatures—so long as they adhered to the standards set by the CCA. As of February 2018, there were five service providers empanelled with the Ministry of IT.

Keep It Safe

But Sharma's mind was still at it. If legally binding documents of all kinds are getting to be a reality, where on earth are you to store them? The most obvious answer to his mind was to build an API for a digital locker—or DigiLocker as it is now called.

Sharma decided to sound out the founders of iSPIRT about the idea. (iSPIRT is the acronym for Indian Software Product Industry Round Table, a think tank founded in 2013 by Sharad Sharma, Bharat Goenka, and Anand Deshpande. The three were part of a vocal minority who believed India must develop a strong software product ecosystem and not stay wedded to the software services ecosystem. With India Stack, they could envision a set of tools to build software out of India, with iSPIRT as a body to evangelize it. They hitched the think tank to the future of India Stack. Sharad Sharma is its most vocal face on all public forums.)

The iSPIRT team immediately suggested Amit Ranjan as an obvious candidate to help build DigiLocker. He was wedded to the open source movement and had earned much respect from peers in Delhi for having organized the first BarCamp, or an Unconference—a loosely structured gathering with a focus on informal exchanges—there against much competition from his counterparts in Bengaluru.

When the call from R.S. Sharma came to him, Ranjan was sitting pretty after having sold SlideShare—an entity he built with his sister Rashmi and brother-in-law Jonathan—to LinkedIn for a few million dollars and was wondering what to do next. He remembers being intimidated by the call and Sharma's officious sounding voice. (Sharma is not the type to expend energies on pleasantries. He asked Ranjan to think about the offer and left it at that.)

'Would anybody in their senses ever consider working with the government!?' he recollects, rolling his eyes with much mirth at a coffee shop in Delhi where we met him. Ranjan, like most people, had heard of Aadhaar. He thought of it as an identity number that had already enrolled in excess of 500 million people. Curiosity got the better of him and he reached out to Sharad Sharma, Varma and Nilekani to see if there may be any merit in it for him. They gave him an outline of what they had in mind and urged him to visit R.S. Sharma personally.

It was only when they met that he figured Sharma knew exactly what he was talking about—and that a digital locker of the kind he described hadn't been built anyplace else in the world. 'So, finally, I made up my mind to push my professional plans to three years later and give this project a chance. I told him, okay sir, I will take this project up,' Ranjan recalls of the time in 2014.

Once in there, he figured why he was needed. While he was surrounded by intelligent bureaucrats, they were generalists. They need specialists as well. 'Those in the government understand that. They had already seen Aadhaar could be implemented. To make it work at scale, APIs needed to be built on top of it. They were honest enough to admit builders from the outside must be engaged. That is why they were talking to me. You've got to appreciate them for that.'

When he got down to work, he was first asked to build a minimum viable product in line with all protocols that were followed to create the digital identity: use open source software, devise architecture that allows multiple digital lockers to coexist, and is interoperable.

The stated intent here, of interoperability, Ranjan recounts, is because R.S. Sharma could imagine a future where it would be possible that different organizations may have their preferred locker to store documents in. For instance, the authority that issues examination mark sheets may choose one locker to upload the documents, while the Regional Transport Office (RTO) may prefer another digital locker service provider.

Ownership of the document, though, must lie with the individual to whom it is issued. This can be established by tagging each document with the Aadhaar number. They can claim a digitally signed version of the document by authenticating their identity against this number and using their biometrics.

Entities can request to view these documents for specific purposes. A university may want to inspect the educational records of a candidate who applies to it for higher studies. Or a bank can ask for the financial records of an applicant who may have applied for a loan. The viewing of these documents, though, is subject to the owner's consent. It is also possible a consumer may want to put all their documents in one locker. That feature is still being worked on.

Most documents that can now be accessed online originate from various government bodies. A few private entities, such as FundsTiger, BankBazaar, IndiaBulls, and BajajAllianz, have registered themselves as entities willing to give up the physical route and take this up if a customer allows them to.

Let Go of Cash

'A lot of [these improvements and advances] also happened because we were at the right place at the right time,' says Varma. But at times like these, people who can see the big picture help. 'If you have come this far, why aren't you thinking of a cashless layer as well?' Nilekani asked Varma one day. 'He was thinking the big picture. Sanjay and I are the foot soldiers,' he says.

The two realized Nilekani had a point. Most Indians had come to take instant money transfer for granted. It could now be transferred at any time of the day, from one bank account to another, to any other person or entity, using a facility called Immediate Payment Service (IMPS), which was launched without much fanfare in 2010 by NPCI.

NPCI had also launched a USSD-based mobile banking service that worked even if somebody used a 'dumb phone' and latched on to a 2G network. It was built for a different era. If somebody dialled *99# on their phone, all basic banking facilities could be availed of, including fund transfer, 24/7,

365 days a year. However, it had very limited reach, with only two telecoms offering it. 'It was like a hidden gem,' says Varma.

Hota points out that most people don't appreciate the fact that India was the second country in the world after the UK to have a system like IMPS in place. And when all of them examined the architecture closely, they figured this could be 'a fun thing' to work on, says Varma. What if they built an API on top of IMPS and made it possible for people to send and collect money as easily as they sent and received an email? There would be no need to remember something like a bank's IFSC (Indian Financial System Code). In any case, conversations began at NPCI on what could be done to make IMPS better. To figure this out, much ground research was done.

Representatives from e-commerce businesses such as Myntra, Flipkart, and Snapdeal, among a few others, were called in to ask for their experience in India. They had a problem with credit cards. There are too few of them in India—and those who do have them are often turned off because they need an OTP to authenticate and complete a transaction. If there was an easier way to make business grow, they'd appreciate it. Technically, this was a case for a better payment system. They hadn't started work on one though.

Varma, Jain, and team got down to work and started to look at payment systems across the world. Among those they looked at closely were Apple Pay in the US and Ali Pay and WeChat Pay in China.[5]

Apple Pay has relationships with almost every bank in North America. iPhone users can buy pretty much everything if the details of their credit cards are stored on their devices. Concerns have been expressed on multiple forums because Apple is a large technology company and uniquely placed to look into how people use their money. Apple was competing with others such as Paypal, Google Wallet, Android Pay, Amazon Pay, etc.

China has Ali Pay. It now has 450 million users, and like Apple Pay, allows them to pay for pretty much anything using their mobile devices. It has created a monolith that percolates into the lives of people in ways that were unthinkable a few years ago. Now housed under the umbrella of an entity with interests that span everything from retail payments to private equity, it has technologies to extract as much intelligence as it can from the data it harvests. China is home to WeChat Pay as well, which is the country's equivalent of WhatsApp. WeChat has over 900 million active users, of which at least 600 million use WeChat Pay to make payments.

Both of these entities are furiously at work to sell people in China a story—that it is fashionable not to carry cash. A survey conducted in 2017 has it that 86% of urban Chinese are comfortable not carrying cash[6] so long as they have access to their mobile device.

Once these trends were studied, it was time for Varma and Jain to brainstorm with NPCI officials to ask if it was possible to re-imagine the future of payments in India. NPCI had done a lot of the heavy-lifting. The RBI was most supportive. The business ecosystem was asking for it. The government could gain. And over all else, the goal of financial inclusion could be met.

There was one thought on everyone's mind though. From what they could see in other parts of the world, those who had gotten into the domain wanted to dominate the space to gather data. How could they prevent that?

What if they built this out as a public commons on top of which services could be built?

Venkatesh Hariharan, a former journalist, value investor, and the director of the Fintech vertical at iSPIRT, offers perspective on how the UPI got to be built as a public good. When the think tank started to look around, it saw a country where cards were just picking up. But the ecosystem stood at an inflection point where the entire payments ecosystem could be re-imagined from multiple perspectives. But much thought had to go into it before that could happen.

That India has millions of mobile phones is well known. WhatsApp, Google, and Amazon could see the merit in integrating with UPI. They had already started testing their systems to enable that. There was much at stake here. In China, they were blocked; in the US, private players were running closed systems; India remained the only big market left in the world.

When looked at from an Indian perspective, the entire ecosystem was using open standards. On paper, anybody can come into it. History has it that no technology firm has been able to stand its ground in the long-term unless it is protected by the state. Basis multiple conversations between the Finance ministry, RBI, NPCI, and representations from iSPIRT, a call was finally taken. Since the standards are open, everything built on UPI must be interoperable.

The floodgates opened, and overnight Indians witnessed three kinds of payment systems.

- NPCI had already created a system that allowed anyone with a regular feature phone to access their bank account.
- Those with a smartphone could use a payment app on UPI. Business entities could see the strategic significance of this. It is the most efficient way to gather data. Among the first

into the fray was PhonePe, a start-up. Paytm wanted a piece of the pie, but was dithering. Flipkart acquired PhonePe. Google launched its own app. WhatsApp has integrated UPI.

- For those without a phone, NPCI launched a service called Aadhaar Pay. To pay, all a customer has to do is enter their Aadhaar number into a biometric-enabled POS terminal that is with the merchant. The bank accounts linked to the number show up. The customer can choose the account to pay from and authenticate it against their biometrics.

When all this is put together, the nature of banking can change completely: cash can virtually be eliminated, and customers can have better visibility on their transactions and use the data for their own benefit. It also alters how banks look at their future, points out Venkatesh.

This, he says, is because an entity can use APIs to build multiple applications. An entity is now limited only by the imagination of its promoters. When the potential futures were presented to banks, entrepreneurs, and those from the start-up ecosystem, the reactions, he says, ranged from denial to great optimism. Who could have imagined that a start-up could emerge out of nowhere, create a payment system, and morph backwards into a bank? For that matter, how was anyone to know this thing called India Stack would get the attention of those in the telecom business? They had built a point of presence deep into India and this is just the kind of thing they needed to get into the financial services business. Banking would have to reinvent itself.

On the ground, as with the government, the private sector also struggled with migrating to the new system. To fraudsters, it became yet another way to dupe customers. JJ College in Koderma, Jharkhand, reportedly lost Rs 1.13 million in an Aadhaar-related scam.[7]

Expand the Economy

Vijay Kelkar has lost much sleep over the fact that at the turn of the century, India had not managed to reform tax structures or start work to build out public goods and services. It is on the back of these concerns that he crafted the proposals for GST when in office as chairman of the 13th Finance Commission. In tax compliant countries, it is easy for the government to fund public goods and create public services because it puts money in the government's hands.

While Kelkar has not worked on India Stack, his intellectual imprint is evident on Nilekani and the designers of India Stack. As Nilekani said in a conversation with Bill Gates on how India Stack would evolve:[8]

> India cannot follow the manufacturing route to growth that Japan, Korea, and the rest of South Asia followed. [This is because] manufacturing itself is becoming highly automated, and there is Chinese overcapacity in every sector, from solar panels to steel. So that is out. Then with this de-globalization feeling around the world like Brexit, Trump, I think global growth won't happen. The golden era is over. India's domestic growth has to be services based.

That there is much merit in these arguments is borne out by the fact that after much public money was spent on building roads in many parts of the world, there is evidence to suggest the incremental benefits that accrue out of spending more[9] is not worth it.

'In public policy, the key question you must ask when thinking about the impact of investing into something is whether it holds the potential to expand the size of the economy; or does it replace something that already exists?' Niranjan Rajadhyaksha, a respected writer on economics, told us. His submission is that 'oftentimes, people jumping for what is new and shiny is perceived as growth.'

By way of perspective, if taxi aggregation services like Uber or Ola get into an economy, the question to ask is, have they created more jobs, or have they replaced old taxis that did the same job? If they only replaced old taxis, the economy did not expand.

Once GST is implemented, in theory, the economy will expand in the long run even if it contracts in the short run, because it holds much potential to eliminate multiple inefficiencies. That it was badly implemented at the outset is another thing. Much of it has to do with infighting between the Finance ministry and bureaucrats from the IRS on who gets the plum positions. By the time a consensus was arrived at, much time had been lost and what was finally offered on the table had too many slabs, was confusing, and ran contrary to all of what Kelkar had recommended. It boomeranged.

Many businesses—especially the small and medium ones—felt the pain of transition to the GST regime. On one hand, they were bogged down by the cumbersome procedures and extensive documentation. On the other, they faced issues with the GSTN portal itself. Built by Infosys (it had won the contract in 2015, competing against Microsoft, among others), it was criticized by users for being too slow and full of bugs. Infosys said that this was because of scale and integration with multiple stakeholders. GST, as a tax system, seemed to be bogged down by complexity.

The Modi government wanted solutions, and it reached out to experts. One of them was Nilekani, who had by this time re-joined Infosys. Complex systems can often be tamed by applying the principle of Occam's razor, which says that the simplest solution is often the best one. And that's what Nilekani did. In a presentation to the council, he said that the system needs to avoid three big mistakes: it should not be burdensome to taxpayers; it should not rely on tax intervention by officials; and it should not have higher levels of mismatch. (GST depends on input credits based on transactions between a vendor and a customer.) He proposed a system that aligned with the business cycle of verification and payment of supplier invoices.

Thomas Isaac, Kerala's Finance minister, and no fan of the central government, was all praise for Nilekani's suggestion. It was 'remarkably simple,' he wrote on his Facebook page. 'My only doubt is, why couldn't the government talk and bring in this chap since the beginning?' In July 2018, the GST Council announced a simpler structure and also said it had accepted Nilekani's model with some tweaks.

It's in the nature of complex, evolving systems that things take place only when certain windows open. Consider the BHIM app in the aftermath of demonetization. Soon after the announcement was made on 8 November 2016 that 500- and 1000-rupee notes were no longer legal tender, chaos ensued as people panicked. Very soon, it was clear to those in the government that they were unprepared for what lay ahead. A quick fix was needed. The team that had worked on India Stack was called in to create a payment app that could be used by all kinds of people.

While NPCI's UPI had about 25 apps running on it, they were built by banks or private entities. Prime Minister Modi could not endorse any of these. Opposition leader Rahul Gandhi had swung into action. Paytm, which had gained much traction during those days, was used as a barb to attack the prime minister—Pay to Modi.

Finally, NPCI was called in to develop an app that Modi could personally promote and reassure citizens all will be well. They were given three weeks to complete the project.

NPCI got in S. Nikhil Kumar, an engineer from iSPIRT, and turned to Nilekani, who was also one of its advisors. 'We went back to basics,' recalls Kumar.

Multiple constraints were imposed.

- The app had to be so simple it could be used even by the untutored.
- It had to be less than 3 MB in size.
- It should seamlessly integrate with all banks and the switch at NPCI without their engineers having to change anything at their end.
- It should be easy for banks to meet all RBI regulations.

When asked how they pulled it off in three weeks, Sharad Sharma shrugs and says, perhaps it may have to do with the fact that it was built in Bengaluru. 'One unique characteristic of this city is that people line up to help you, especially when they know you have taken up a big challenge.' An entity called Juspay deployed their best people to develop the app and worked 24/7. Another start-up called Reverie helped with translations. DailyHunt, a news app, did the proofreading. 99 Tests, a crowd-testing start-up, did the testing. PhonePe, with which this payment app would compete, offered them the use of its FAQs and other literature.

On 30 December 2016, at a function organized in Delhi, Modi launched the BHIM app by buying a shawl from Khadi Bhandar, and 'paid' digitally. That it was done offline was not known to those covering the event. Not because BHIM wasn't usable. But because the bureaucrats were petrified if it failed for any reason with live cameras from across the country whirring, people would get angrier. Over the next few days, BHIM saw thousands of downloads and is the most used app for digital payments.

At the end of it all, sometimes by design, sometimes by propitious circumstances, and sometimes triggered by a crisis, India has a strong and evolving digital public infrastructure. This is quietly turning the country into a digital economy that produces a large quantum of data. Meanwhile, a consent layer (billed Data Empowerment and Protection Architecture) that allows people to consciously share their own data for their benefit is being built.

Now that all this has been done, India Stack opens many doors.

India Stack Door #1

Data unleashed by smartphones, the GSTN, and digital payments, passing through the consent layer of India Stack promises to fundamentally alter the nature of banking and finance. Imagine a small enterprise that supplies office stationery to a well-known, large entity. The contract between the two stipulates that the vendor supplies on the 1st of each month; and that an invoice be raised on the 10th of each month. Typically, payments are released only 60 days later.

As far as the small vendor goes, to stay in business, 60 days is a long time. But that is the nature of business, and it needs capital as well.

With digital payments, GSTN, etc., kicking in, to stay afloat until those payments come in, the vendor can take digital records to a potential lender, show them future cash flows, and use the expected cash flow as a collateral to borrow for the present—in technical parlance, opt for flow-based lending.

In developed economies, flow-based lending from financial institutions can be as high as 65% of its business because it has complete visibility on a borrower's credibility and what their cash flows look like. In India, it currently stands at 8%.

This potential has been spotted by some early entrepreneurs. They have banded together[10] and have formed the Digital Lending Association of India to represent their voices. Venture capital has started trickling its way into this space as well. This opens up exciting possibilities. As cash-flow-based lending starts to grow, it opens room for much-needed capital to get into the system, stoke entrepreneurship, foster innovation, and create jobs.

Venkatesh Hariharan's research suggests that this means multiple things. Currently, only 7% of small and medium enterprises (SMEs) in India have access to formal credit from the banking system. The rest rely on their friends and family network. Then there are at least 100 million households in India that cannot access formal credit in spite of their annual income being in the region of Rs 200,000–250,000.

This, to his mind, has much to do with banks not having access to data. Some states like Andhra Pradesh seem to have gotten the import of what this means and the more nimble banks like Yes Bank and Ratnakar Bank Limited have cottoned on to the potential of India Stack. Between a state government like Andhra Pradesh and these challenger banks, there is much work that is happening to craft a narrative for flow-based lending.

Varma and Jain say this was a future they hadn't imagined in all its details when they'd brainstormed on this. What was at the top of their minds was how to make money transfers for the government for its various DBT schemes. NPCI had done some of the basic work on how money must be transferred in India. UPI was built on top of that. And the innovations are happening elsewhere, in an altogether different context. Who knows what else may emerge? That is the nature of business and entrepreneurs.

India Stack Door #2

One of the biggest grouses entrepreneurs have against the country is the lack of infrastructure, which impedes movement of goods from one state to another. To build it will take time. But a door has certainly opened.

With GST and GSTN in place, the GST Council approved the e-Way Bill. Behind the scenes, multiple glitches in the system needed sorting. The idea is that the system will allow a transporter to load a truck at, say, Amritsar in North India and drive it to Kanyakumari in South India without stopping anyplace or carrying any cash.

Until very recently, each state in India had its own tax laws. With GST in place, India is now looking at a 'One Nation, One Tax' future. Earlier, if farmers in Punjab had to transport their produce to Tamil Nadu by road, the trucks would have to stop at various points in the states they passed through.

On average, trucks spent 20–30 minutes at checkpoints in Rajasthan and Maharashtra, and up to two hours in Bihar or Jharkhand. But with the e-Way Bill, all checkpoints will be dismantled. This clears waiting time right away.

A World Bank study[II] has it that logistics account for 14% of the total value of goods in India. In a developed economy this is in the region of 6–8%. To bring these costs down, much work is being done to create dedicated freight corridors. But that isn't enough. Studies have found that up to 60% of the time, a truck does not move. Instead, it is stuck at checkpoints to pay tolls, taxes, and clearances to move from one state to another. To get around these, truckers from places like the textile hub of Tirupur in Tamil Nadu divert their vehicles a few hundred kilometres to avoid the Tamil Nadu–Kerala border crossing. In turn, this pushes costs up significantly. With this bill in place, such costs can be slashed and India can become a more competitive place to do business in.

Presentations made to the departments have argued this move alone will reduce diesel consumption by 7%. This also suggests that it will make India a more viable place for e-commerce companies. Currently, 50% of all e-commerce transactions in India are conducted physically. But there is a cost involved to handling cash.

If these efficiencies kick in, Flipkart, for instance, which was acquired by Walmart for $16 billion in May 2018, will earn, on average, Rs 30 more per transaction on such orders. It is much the same thing with Amazon. These savings, as arguments go, can be deployed by entities such as these to set up fulfilment centres or acquire other entities to expand their businesses.

India Stack Door #3

India Stack can be deployed by an entrepreneur to build something completely different as well. In the healthcare space for instance, there are many things that can be explored. This is what has got the Gates's attention and that of their Foundation, where much of their time and personal wealth are spent. Granular studies have revealed much. Take one example as a case in point: The percentage of women who deliver babies via C-section in Shravasti district of Uttar Pradesh is 0.5%, often because they do not have the resources to go to a hospital. In Europe and the US, at least 25% of babies are delivered by C-section. Because they don't go for it even when they have to, the chances of babies born in this district dying or suffering serious deformity are high.

On the face of it, this is a problem that can be fixed. If an organization builds a system to deliver funds directly to a person's account so they can pay for basic healthcare when most needed, evidence suggests it will bring down infant mortality rates. At least in theory, what it means is that if Project Aadhaar is being thought of as a digital public good, the next stage should lead to the creation of a digital public service as well.

It is much the same thing with access to basic vaccines and diseases like tuberculosis that have been wiped out from the developed world. That India hasn't managed to eradicate such diseases isn't for a lack of funds, but because of flawed design principles. By way of example, in Union Budget 2018, Finance Minister Arun Jaitley announced the 'world's largest government-funded healthcare programme'.[12] He announced it will provide secondary and tertiary healthcare cover to at least 100 million poor families and each family would be covered up to Rs 0.5 million.

But here again, those who were involved in the deliberations argued this is uncalled for. What these 100 million families really need is access to primary healthcare facilities and up to Rs 30,000 by way of cover. Those familiar with insurance know this proposition is more expensive to service because people need—and will use—primary healthcare over secondary and tertiary care.

It was argued that here again, the many elements of India Stack could be one of the multiple ways to ensure the system isn't pilfered with and the money reaches the intended beneficiaries. But to the masses, Rs 500,000 sounds heck of a lot more than Rs 30,000—and politicians know tools like these are called for when elections are around the corner.

It is a thought policymakers at the highest echelons have been sounded off on—that morality not be sacrificed at the altar of expediency. There is much deliberation around it now.

India Stack Door #4

Then there is the fact that this being a public good, it has been built as an open 'stack' as opposed to a 'closed' platform. The difference between the two is an important one.

Facebook, Google, and Amazon, for instance, are platforms for business. But they are closed ecosystems. Anything that is built on these platforms has to be built using protocols these entities think appropriate. They set the standards, own the user data, and can extract value out of them.

As opposed to that, a 'stack' is a set of APIs that have been bunched together and are intended to foster innovation. The standards are open. There are no restrictions on what is built using this stack of APIs. It just happens to be that what exists on it now is built for India—that is why it is called India Stack. It can be offered by the Indian government to reach out to some other country—and they may want to call it something else.

India Stack Door #5

In the second week of February 2018, a scam of unprecedented proportions dominated the headlines and all discourses across India. A billionaire diamond merchant called Nirav Modi had scammed the state-owned Punjab National Bank (PNB) of Rs 110 billion.

There was much sound and fury because Prime Minister Modi did not say much around what is without a doubt the largest heist in contemporary Indian business history. If it were to happen in any other part of the world, heads would have rolled overnight.

At the time, when probed on why exactly an otherwise vocal prime minister had stayed quiet thus far, an individual who interacts with key decision makers at the PMO declined to take calls. Instead, he responded with a text message: 'We are still developing our perspective.'

It turned out that the PMO was looking at the scam from an IT perspective. One of the reasons why the scam—which centred on the issuance of letters of undertaking (LoUs)—went undetected for a long time within PNB had a lot to do with its systems not talking with each other. Its SWIFT (Society of Worldwide Interbank Financial Telecommunication) messaging system was not connected to its core banking system, and there were indications that unauthorised personnel were issuing LoUs using the system.

The concern was that PNB was not the only one that was struggling with disconnected IT systems. Many other state-owned banks faced the same issue.

The PMO was quick to see the broader picture. At a national level, disconnect between different systems was not only slowing down the economy, it was also leading to tax evasion, fraud, and leakages. The PMO was impressed by Nilekani's approach to solving the complexities within GST, and saw that India Stack was also trying to solve a similar problem using platforms and APIs to connect disparate systems to stimulate innovation. The PMO was also aware of the risks of connecting different systems as well as those posed by poor implementation.

The PMO wanted the government to learn from the experience. It held extensive conversations with technologists. It requested Rajiv Mehrishi, CAG of India, who also serves as chairman of the United Nations Board of Auditors, to spend at least 10 days in Bengaluru from the first week of March 2018. His agenda was to examine the architecture of India Stack closely and see how it can be applied elsewhere.

Sharad Sharma of iSPIRT believes it will work at scale across sectors. 'Why just banks, we can reduce medical insurance-related hospital frauds as well. India Stack helps get there by turning things paperless and by turning cash trails into digital data. India Stack is an enabler for applying this idea at scale,' he told us.

Who knows what else follows when a butterfly flaps its wings?

Notes

1. Peter Dizikes, 'When the Butterfly Effect Took Flight', *MIT Technology Review*, 22 February 2011, https://www.technologyreview.com/s/422809/when-the-butterfly-effect-took-flight/, viewed on 7 July 2018.
2. Ministry of Finance, 'Pradhan Mantri Jan Dhan Yojana', Government of India, https://pmjdy.gov.in/scheme, viewed on 7 July 2018.
3. Ministry of Electronics and Information Technology, 'Digital Signature Certifi cates', http://meity.gov.in/content/digital-signature-certificates, viewed on 7 July 2018.
4. Ministry of Electronics and Information Technology, 'Digital Signature Certificates', http://meity.gov.in/content/digital-signature-certificates, viewed on 7 July 2018.
5. Arjun Kharpal, 'Chinese Internet Giant Tencent Launches WeChat Pay in Europe to Challenge Alibaba's Alipay', Tech Transformers, CNBC Special Report, 10 July 2017, https://www.cnbc.com/2017/07/10/wechat-pay-europe-launch-tencent-to-challenge-alipay.html, viewed on 8 July 2018.
6. Masha Borak, 'The Rise of China's Cashless Society: Mobile Payments Trend 2017', *Tech Node*, 15 August 2017, https://technode.com/2017/08/15/the-rise-of-chinas-cashless-society-mobile-payment-trends-in-2017/, viewed on 8 July 2018.
7. A.S.R.P. Mukesh, 'Aadhaar Seeding Scam Takes Root', *Telegraph*, 2 September 2017, https://www.telegraphindia.com/1170902/jsp/frontpage/story_170423.jsp, viewed on 18 August 2018.
8. World Tech Founders Podcast, 'Bill Gates and Nandan Nilekani on India's Digital Transformation', *Financial Times*, 28 November, 2016, https://www.ft.com/content/91c146af-24f0-418b-b5ab-4020bf96aa6c, viewed on 8 July 2018.
9. Marlon G. Boarnet, 'New Highways & Economic Growth: Rethinking the Link', *Access Magazine*, https://www.accessmagazine.org/fall-1995/new-highways-economic-growth-rethinking-the-link/, viewed on 8 July 2018.

10. Venkatesh Hariharan, 'The Rising Tide of Cash Flow-Based Loans', *Mint*, 29 September 2017, https://www.livemint.com/Technology/YThya7EVFnCgphQY1ow3RK/The-rising-tide-of-cash-flowbased-loans.html, viewed on 8 July 2018.
11. Sebastian Saez and Arnab Bandyopadhyay, 'Connecting India's States with Good Logistics', World Bank, 15 February 2017, http://www.worldbank.org/en/news/opinion/2017/02/15/connecting-indias-states-with-good-logistics, viewed on 8 July 2018.
12. Abantika Ghosh and George Mathew, 'Union Budget 2018: Rs 5 Lakh Health Cover for 10 Crore Poor, Prescription Awaited', *Indian Express*, 2 February 2018, http://indianexpress.com/article/business/budget/union-budget-2018-arun-jaitley-national-health-protection-scheme-health-cover-narendra-modi-5048576/, viewed on 8 July 2018.

5

Insurgents, Incumbents, Pioneers, and Leaders

WITH AADHAAR IN PLACE, NEARLY 1.2 BILLION PEOPLE now have an identity.[1] With India Stack, their integration into the formal economy has begun, and a digital ecosystem is evolving. It holds the promise to resolve multiple problems and make life easier for citizens, businesses, and society at large.

For instance, access to finance is a problem for most Indians. This, because most Indians cannot authenticate who they are in the format prescribed by the regulatory authorities. With no money in hand, it is not possible to move up the economic ladder.

For institutions like banks that have money to lend, the issue is authenticating identity—how do they know for sure whom are they lending to? Until now, they have not had access to reliable data for large numbers of people. This restricts the universe of people they can lend to, and impedes their growth.

Between both these problems, economic activities of all kinds suffer, and a country is condemned to stay poor.

With Aadhaar and India Stack, the problem has turned on its head. Almost everybody now has an identity, many are consuming data, and creating a data trail that can be audited. 'It is counter-intuitive,' says Professor Mohanbir Sawhney, of the Kellogg School of Management. He is a management consultant and sits on the board of many global companies. 'But the fact is, Indians are now among the largest consumers of data in the world because they have no other options to entertain themselves other than their phones.'

Data consumption of this kind has created an altogether unique situation. Able to prove their identity, large masses of people can now get into the formal economy, access credit from formal systems like banks, and start imagining going up the economic ladder. Yet, there is no taking away from the fact that while they can be identified, most of them are poor. Businesses must think up new ways to earn profits by serving them. That a fortune awaits those who can serve the bottom of the pyramid is now acknowledged wisdom.

'It is all about volumes and is an old idea called "sachet marketing",' says Sawhney. 'Sell something for Rs 10 to 10,000 people.' This was an idea pioneered in India by Hindustan Unilever when it introduced sachets with shampoos that could be sold for Re 1 each to those who may otherwise not be able to afford it. The idea was fine-tuned in India until Unilever could make its products affordable to two-thirds of the global population that earn less than $1,500 each year.

Mary Meeker, an acclaimed Internet analyst and partner at the venture capital firm Kleiner Perkins, offered some pointers to why this matters in the contemporary world at the Code Conference in California on 31 May 2017.[2] Representatives from governments across the world, global bankers, and investors of consequence were listening in:

> With a digital identity, people can now activate a mobile phone in 15 minutes; once upon a time, this could take up to three

days. Start-ups such as Paytm, MobiKwik, Oxigen Wallet, Citrus Wallet and ItzCash are wooing those with cell phones to create digital wallets into which money can be deposited—large banks are wooing them as well to open accounts. This, because the government is depositing pensions and benefits from other social schemes directly into the accounts of the rightful recipients.

Meeker's analysis is that since Aadhaar and India Stack started to be implemented, the pension and social services payout to the average Indian worker has gone up by 12–15%. They are getting richer and gaining spending power.

Those with spending power and access to the Internet via devices like the cell phone are, in turn, using it to augment their education. By way of example, there are 250 million children enrolled in the K-12 system. The system isn't good enough for them. That is why they are using apps developed by private entities like Byju's. Studies have it, Meeker pointed out, that learning outcomes have improved by 15%.

Much like education, Meeker described the emergence of other start-ups in the financial services sector, healthcare, media, and agriculture.

Her point is, people are beginning to spend time and money in these areas. This, she says, is because the number of people who are part of the 'consumption class' is rising. From 14 million in 2005, 66 million households had moved up the ladder by 2015. Those in the 'consumption class' are people with income to spend after meeting their basic necessities. The number of these households will increase and grow faster as India Stack goes deeper, she predicted. It holds much potential to shift the trajectory of the Indian economy.

It is time then, she said, to pay attention to the emerging India story and understand what policymakers are thinking and how are entrepreneurs exploring the many opportunities.

What Meeker was presenting to the world had much to do with the deliberations of the Nachiket Mor Committee Report that was presented to the RBI in 2014. Members on the committee had gone back to first principles and asked: What is the purpose of a bank? The answer: Lend money to people who may need it and earn profits by way of interest.

To do that, banks must seek out places where people need money the most, lend, and earn as much interest as they can. Evidence has it, though, that banks are not going to every corner of India to lend to those who need money. One member on the committee who did not want to be identified said: 'Banking in India is not really about making money. People who work at these institutions are the elite. They have a general distaste for the poor. Because they smell dirty and they look different.'

This prompted them to ask another question: Why ought banks be allowed to retain a monopoly on the business of money? Why not permit those who may be amenable to deal with those who 'smell dirty and look different'?

When questions on how to identify them came about, those on the committee could see Aadhaar as a tool at a national level that could be accepted by the regulatory authorities. If deployed thoughtfully, banking can be opened up to entrepreneurs who may not have a problem doing business with those that banks looked down upon.

The outcome of all these deliberations was a proposal in the report that an altogether new structure called Payment Banks be permitted to operate. These entities, the committee proposed, should be allowed to engage in all activities a bank does, except lend money. No precedents existed elsewhere in the world.

This structure though, the report argued, is necessary. Banks have no incentive to do business with the poor. This has much to do with the cost of identifying an individual. There is a cost to service them as well. The poor, though, have little money to park as deposits or to pay the transaction fees for any services they may avail from a bank.

Data has it that close to 90% of all deposits in Indian banks are below Rs 100,000. Bankers like larger numbers. If they go closer to the poor, this average may drop even lower, and the cost to service them will rise higher. The poor, to banks, are fundamentally an unprofitable enterprise. While India Stack can drive all costs down, traditional banks will take a long while to wrap their head around the idea.

Then there were voices such as T.V. Mohandas Pai, former board member at Infosys and part of the Vijay Kelkar Committee on tax reforms. They were arguing that banks that service 100 million customers with less than 100 people on the rolls are able to do so because the technology now exists on mobile phones. For instance, it is possible to remit money using a mobile phone to somebody who may need it back home. Or they may want to save a small part of the meagre income that accrues to them each day in a fixed deposit. Those at the RBI were listening to them as well.

Eventually, the central bank conceded, and permitted an altogether new class of entities called Payments Banks[3] to start operations. Those with capital and no exposure to the financial services sector were among the first to express their interest in the idea. Eleven entities were granted permissions.

Since the time they went live, an ecosystem has emerged that allows Indians to authenticate their identity digitally, submit documents via eKYC, and open a bank account without being made to feel dirty and unwanted.

In short, the Nachiket Mor Committee suggested that Indian regulatory authorities unbundle banks. Why, it asked, ought banks be the sole arbitrators of all things to do with money? That was only the beginning. 'There is a long way to go before banks are completely unbundled. This recommendation was a first step in that direction,' says Bindu Ananth, co-founder of Dvara Trust, which helps the poor get access to finance.

Mor credits her for doing much of the ground work and crafting the proposals.

The Insurgents

In 2000, a young college boy set up www.indiasite.com to push cricket scores and jokes over the Internet. Two years later, it was sold for $1 million. Fifteen years and many millions later, Vijay Shekhar Sharma was among the first to bid when the RBI formally gave the go-ahead to payments banks.

With a licence in hand, he could see the door open to gain entry into the ecosystem that is India Stack. He had founded Paytm, a digital wallet, in 2010. If he got people hooked there, he figured it could eventually morph into a platform of consequence and much money could be made.

In the short term, he was betting that if he could lure people from the middle and bottom of the pyramid to use Paytm, money deposited into the wallet would stay parked in an escrow account. Until people used it, he could earn money off the deposits—in financial parlance, the float.

In the longer term, he knew that the more people used his wallet, the more data he would get about them. How to deploy it is where he applied his imagination—and bet the company's future on. Those who have worked with him say he is the kind of person who can take punts that only somebody who understands India deeply can.

Sharma grew up in Aligarh, far away from the metropolitan cities of India. That may be why he could think up an idea that made no sense to anybody else in its early days. A few hundred million people take the bus every day to travel from one city to another. Paytm made an offer: Use money from the wallet to book bus tickets, and get the trip insured for free. If they changed their mind up to six hours before the trip, the money would be refunded to use later.

This cancellation policy was an insurance product of the kind that didn't exist in the market. While other insurers tried to wrap their heads around how to beat this one, Paytm was gathering data on when Middle India and those at the bottom of the pyramid choose to travel most. They now know it is between Monday and Wednesday.

Basis multiple experiments like this, Paytm has arrived at a playbook of sorts.

- Stay focused on the middle and bottom of the pyramid.
- The most important metric to measure is the number of users on the platform. That means, 1,000 less active users are better than 100 active users. Because even less active users bring data with them.
- Deploy every trick in the book to get people to transact. It is better to have 100 users conduct some form of micro transaction as opposed to 10 users who may choose to spend large sums of money. The more people transact, the more insights you gain into what is on their mind. Data matters.

Paytm was growing at a manic pace.

What Sharma hadn't reckoned with was that the RBI would mandate that wallets be made interoperable—much like money can be transferred between bank accounts. And that Google would launch Tez, and WhatsApp would integrate UPI as well.

'In India, the customer at the bottom of the pyramid is very fickle. To create a brand there takes much time and requires a lot of patient capital. With interoperability, the cost of switching from one wallet to another is zero. Now there is nothing to stop anyone from moving to Paytm to WhatsApp or Jio or, for that matter, a regular bank. What it boils down to now is, how much is Paytm willing to pay to retain a customer?' says an angel investor who has been watching the entity with much interest.

By all accounts, Sharma was upset after WhatsApp decided to get into payments by integrating UPI. It was evident to him only then that he had held on to the wallets business for too long. He stepped on the gas and integrated Paytm with UPI in November 2017, a couple of months before WhatsApp did.

As in the case of Airtel Payments Bank (mentioned in the previous chapter), Paytm might have crossed the line. On 31 July 2018, the RBI asked Paytm to stop adding new customers because it had concerns about the process it followed in acquiring new customers.[4]

But in growing this far, Paytm has caught the attention of marquee investors like Alibaba from China and SoftBank from Japan, and is now valued at $10 billion[5] and raised $2.5 billion[6] so far. It has outgrown Sharma. What may its grand plan be? Nobody seems to know.

There are other narratives as well around India Stack. This includes Bala Parthasarthy of MoneyTap, and people who fund them, like Sanjay Swamy of Prime Venture Partners.

The problem, as they see it, is that most bankers have been taken off-guard by the emergence of Aadhaar and India Stack.

With the UPI component of India Stack in place, the fees-based income for banks have plummeted. What they are left with is the lending business. This has two parts.

- Personal loans, collateralized loans, and micro-finance loans are the ones most people are familiar with.
- Cash-flow based loans have everyone's attention. They need expertise and data, though. If there's a way to figure what a borrower's future cash flow looks like, it can be used as a collateral to lend money.

Convention has it that before money be lent, two questions be asked: Does a borrower have the ability to repay? To answer that, a lending institution must first figure out a potential borrower's current earnings and liabilities. This tells a bank whether they may have enough money left to repay a loan it offers. Trying to find an answer to this question is expensive. On average, a bank must lend Rs 500,000 to an individual to earn any money off them. The most efficient of these is HDFC Bank, which can earn money by offering personal loans in the region of Rs 400,000.

The ability to repay is one thing. But does the borrower have the *intent* to repay? There are no models that can predict intent. Attempting to answer this question is a tough one. Parthasarthy's MoneyTap is one of the many start-ups in India engaged with the question. To grapple with the question though, formidable algorithms and technologies must be built. The first opportunity he spotted in wrestling with the question has got the attention of entities like Aditya Birla Finance Limited and Ratnakar Bank Limited.

Getting to the next one millionth customer is a very expensive proposition for an Indian banker. That is where technology can make a difference. If Rs 65,000 be the average limit on any card a bank issues, it needs at least $1 billion at any given point in time to reach out to 1 million people. 'It's a huge amount of money,' says Parthasarthy. So MoneyTap thought up an altogether different product.

After interviewing those who need access to credit the most, Parthasarathy figured Indians don't need large amounts of money. Their needs are very rudimentary. For instance, they may need a small loan to visit their home town, or to meet some expense that may have come up. The ticket sizes here are in the region of Rs 3,000–20,000. Not surprisingly, people feel humiliated when rejected by banks.

As for banks, the economics don't permit them to service people with such low requirements. But with online authentication and eKYC, they can identify these customers and regulatory requirements can be complied with at a very low cost. Some thinking through later, Parthasarathy came up with a solution that works around the issue on hand: as opposed to a big-ticket loan, offer those with an identity, a credit line of up to Rs 500,000. People pay interest only on the amount they access from the credit line they are offered.

It allows banks to invest in places they could otherwise not afford to and brings customers into the formal economy.

For banks, this data serves to build a credit history it can use in the longer term. Everybody within the system intuitively knows there is an appetite for short-tenure loans. Most of it is currently controlled by the informal market. Banks haven't attempted getting to these markets yet—to offer such loans, identification systems were needed and mechanisms to collect money as well.

In these informal markets, a street vendor can take a loan in the morning and retire it in the evening by pledging a percentage of his daily earnings. If they don't repay, the informal market has its own dubious ways to recover it.

With Aadhaar and India Stack in place, work is on to implement a system that allows formal banks to offer short-tenure loans. The intent is to weed the informal economy out and build a tax-compliant society. It is in the government's interest as well. But to do this, formal lenders like banks must feel comfortable. For this to come about, they insist that all risks that surround loans be monitored and controlled.

While the construct is already in place, it will start rolling out formally towards the second half of 2018, when UPI 2.0 begins to get implemented. Version 1.0 of UPI can only lend. In Version 2.0, a lender can digitally recover an agreed-upon percentage of a vendor's daily earning as repayment towards a loan.

The upside to a street-vendor is that he or she can access loans from a bank. The lender gains the confidence to offer credit because they have the first right to an agreed-upon percentage of the vendor's daily revenue. The borrower cannot put this away as discretionary because there is a built-in mandate that automates the system.

'It is the same idea as pre-paid cheques. But here it guarantees the lender it will not go to the bottom of the pile, but stays where it is, and as agreed upon, without any intervention,' says Sharad Sharma of iSPIRT.

To understand how this works, imagine a merchant who sells a certain product at an online marketplace like Amazon or Flipkart. Each month, they may sell anywhere between 50 and

100 units. For a lender, that is a huge variation. To mitigate the risk of offering a loan, if needed, the lender can ask for data on what kind of sales the vendor did in the previous month. This may show the vendor sold 70 units.

To be on the safe side, and depending on their risk appetite, the lender can choose to work on the assumption that the vendor will sell 60 units and offer them a line of credit to buy that many number of units for a month. The technology now exists to monitor the business in real time and on a daily basis. If the vendor repays on time, the lender can choose to renew the loan for another month.

In this kind of a situation, the lender's money is insulated in two ways: Basis data from the past, and after having factored in for an element of risk, it can offer a short tenure loan it reckons is just right. If any alarm bells are triggered, with the UPI mandate in place, the lender has legal rights to take away whatever earnings were due to the institution first. This was not possible in the past.

For instance, an autorickshaw driver, who was not part of the formal economy until now, can now access micro-credit in a formal setting. Dynamic monitoring of the driver's revenues allows a bank to keep tabs and see whether the borrower is on target to earn as much as they did in the previous month. If the algorithms suggest all is well, the tenure of the loan can be extended. Else, it can decline to extend the tenure. But just a mandate alone won't help. A credit history is needed as well.

That is why technology of the kind MoneyTap is building is much needed.

There are entities like Capital Float as well that have gotten into the fray. Their promise is to offer MSMEs and SMEs instant credit using India Stack, which identifies and authenticates eligible entities through Aadhaar and eKYC. eSign is deployed to get the documents signed. Once this is in place, these entities offer first-time borrowers an unsecured business loan for 24–27% interest.

In the developed world, up to 60% of entities access such loans. In India, only 8% of MSMEs have access to such systems, because most of them operate outside the formal economy.

Can this idea be extrapolated in some form to the masses who have established their identity with Aadhaar, but have no credit history? Is it possible to find large, lendable masses of people? Is it possible for any entity with the funds to afford to take a risk by offering a day's loan, or a week's loan? India Stack is architected such that it can be used to intervene here.

If it is done, over time, an authenticated database of people with credit histories can be built. So, assume a street vendor needs a small amount of money to finance his daily business. A bank can now offer it against his or her projected cash flow. If he or she pays up on time, it will have data about one lending cycle. If they do it for 10 cycles, it offers the bank a window to build a database of the street vendor's behaviour pattern.

The algorithms at the backend can then suggest that the vendor now has a good enough credit history to be extended a longer-term loan. The bank can choose to extend his or her credit cycle by a week. Another cycle can be initiated, and if all criteria are met, the tenure can be extended further. Multiple permutations and combinations are possible.

Venkatesh Hariharan of iSPIRT points out the many believers in this model. He suggests Niyogin Fintech be considered as a case in point. It was founded by Amit Rajpal, who used to head the Asia business of Marshall Wace, a hedge fund. Rajpal used to manage $5 billion for the entity out of Hong Kong. 'I have met him, and Amit thinks the India opportunity is big.' Rajpal's pitches have raised $36 million until now for Niyogin. The capital that people like Rajpal raise is what Venkatesh calls 'informed capital'.

Hariharan's point is that people like Rajpal know it takes time to create a good database, and that technology needs to understand and learn about the patterns in data. It is only after much toil that models of any consequence that work can be built. If done well, anything is possible. What's to stop a new-age non-banking financial company powered by technology to emerge, he asks. Having said that, he is quick to remind us that building entities of the kind Rajpal has set out to, takes money and patience.

Evidence from other parts of the world indicates that disruptive innovation comes from new players. Whether this will happen in India is something nobody is willing to bet on yet. This is because all transitions in the past have followed a certain path. Before a country could become economically rich, it had systems built to generate wealth.

In India though, data is being generated out of systems that are still evolving. And its citizens are not wealthy yet. How is an entrepreneur to build a viable business model then?

There is only one thing everybody agrees on: To create anything of consequence in India, products must be created for people who earn below Rs 30,000. And the fintech business is a good place to experiment with models of all kinds because it seems to be amenable to disruption. Because, as Parthasarthy puts it, 'There are no new ideas in finance. Only ways to repackage money. That is how it has been since the time it was invented. The innovation is in the repackaging and the disruptor is the one who does it first.'

This world view is endorsed by a Bengaluru-based venture capitalist who has been watching the Aadhaar story for a long while now and has bet some monies in a few early-stage companies as well. He offers some interesting insights, but asks that his identity be masked.

Some things in economies such as India's can work well only if the transaction costs are low, he says. The reason a formal lending and credit ecosystem has not taken off here is because the cost of identifying somebody and managing a transaction is huge. Reduce that cost and provide fool-proof identity, and everybody becomes 'lendable'.

People who get this include Sanjiv Bajaj of Bajaj Finserv, whom the venture capitalist described as somebody who has a 'visceral belief' in the Bharat Market. 'He [Bajaj] once told me the only part of India that believes in the Bharat market is Bollywood. If a movie works in Bharat as opposed to India, they make money. Bollywood knows how to chase a market. That is something we must learn as well.' His submission is that Bajaj Finserv is among those Indian entities that has figured there is merit in expending time and energy on India Stack.

'Most Indian start-ups,' he continues, 'are weakly positioned because they do not know how to deal with Bharat. What it has now are entrepreneurs like Kunal Bahl, founder of Snapdeal, and Sachin Bansal and Binny Bansal, the founders of Flipkart, who overestimated the size of the Indian market they are familiar with. I'd put entities like Kalaari Capital and Accel Partners as well into this group. This bunch understands New York better than Tumkuru [a district in Karnataka].' When pressed on why, his argument is that in his multiple engagements with them, they are comfortable dealing with technologies on offer by Google and Amazon as opposed to what is being built for an emerging ecosystem. It places them at a natural disadvantage.

Then, he says, there are the American multinational companies. These were built by founders who had to deal with a very messy infrastructure to begin with. 'Think of the microprocessor business, for instance. The battles fought in the US, and American companies have a competitive advantage.' Here again, his point is that American MNCs have a knack for identifying emergent technologies and get their hands dirty to create something of consequence. They are the kind of people who will look closely at India Stack and tinker with it if they see merit in it.

The Challenger

It is for this reason that imagining Mukesh Ambani's Reliance Industries as a traditional conglomerate would be erroneous. 'It operates like a start-up that can spin on a dime,' says Mohanbir Sawhney who is a board member of Reliance Jio as well.

The entity has demonstrated it can disrupt. The most recent example being its entry into the telecom business. Technology companies call the shots in every part of the world. The models they work off, though, are built in a world where purchasing power exists. What is well understood within Reliance is that purchasing power does not exist in India. And that there are profits to be earned if services can be offered to those in Middle India and at the bottom of the pyramid. What may it take to get there and collect small amounts of money that add up to billions?

When deliberated upon, telecom looked just right. This business has the potential to reach at least 500 million people if sold right. When the nature of the business was studied from Jio's boardroom, all telecom companies used 2G and 3G spectrum to deliver voice. The 4G spectrum was used to deliver data. Getting these networks to talk to each other seamlessly is inefficient and expensive. The costs are loaded as tariffs to the consumer and make cellular services expensive.

So they decided to start a telecom company that works exclusively off the latest 4G technology. When the math was done, they figured an hour-long call would consume just about 45 MB of data.[7] But give people access to free data, and they will use it to access content and services of all kinds. This could give the company insights into how individuals behave, and equip it with huge databases.

The company formally launched its services in September 2016. Voice calls were offered free and data was priced at Rs 50 per GB—all bundled into an inexpensive smartphone.[8] This move was a much scrutinized and controversial one.[9] The incumbent telecom players considered Jio pricing predatory. Besides, the telecom regulator had cut interconnect usage charges, which they said favoured only Jio. Challenges against its entry were mounted in the courts, and R.S. Sharma, who was now at the helm of the Telecom Regulatory Authority of India (TRAI), was caught in a legal bind. He sought the counsel of the Attorney General of India (AGI) and the case was scrutinized closely. Many consultations later, Sharma concluded that Jio was not overstepping any boundaries, and green-lighted the project.

On launch day, as anticipated, people started lining up in the thousands to get a device—some wanted multiple devices—with active connections on it. The team was bogged down by the

KYC rules and public sentiment was turning overwhelmingly negative. Accounts from people who were there on launch day suggest that it got Mukesh Ambani all worked up. (This was one of his biggest bets.) Until a suggestion percolated through—that TRAI be asked if it would accept eKYC, that layer of India Stack which permits digital identification. TRAI agreed. Ambani issued instructions that machines be procured overnight from every corner of the country. He monitored the situation on the ground until reports started to stream in that all was falling into place. Customers could leave a Jio store within 15 minutes with a fully functioning phone.

'So, if you think about it, had the government not created Aadhaar, we could not have done eKYC and scaled Jio. And by the way, Aadhaar is just the starting point,' says Sawhney.

One insight the company has is that people consume massive amounts of content on their devices. It may sound evident in hindsight, but this learning emerged from Jio's field trials and multiple brainstorming sessions. That is why it is acquiring content producers. The plan is to eventually monetize this by allowing people to acquire small 'sachets' of content tailored to meet the demands of individuals on the back of data it has acquired.

It has plans to leverage these insights to create multiple streams of businesses in areas as diverse as healthcare, financial services, education, and agriculture.

The Incumbent

Arundhati Bhattacharya, the chairman of SBI until October 2017, was watching India Stack with much interest as well.

This had much to do with the days when the potential uses of Aadhaar was being investigated. SBI was one of the testbeds because it was the implementing agency for subsidies the government offered on gas cylinders. The government of India is the largest stakeholder in SBI.

As part of a pilot, the government issued a diktat that subsidies be de-linked from the selling price and that it be credited to bank accounts instead. When the announcement was made, she says 150 million people used to avail subsidies. Overnight, the number of people went down to 120 million.

Veterans in government know there are the ingenious kinds who will attempt to subvert the system and create multiple accounts. That is why they decided to link Aadhaar numbers to bank accounts that avail subsidies, in addition to other reforms. The number went down to 90 million.

'I am reasonably sure if subsidies are linked to the economic status of a person, the number will go down to 50 million,' says Bhattacharya.

This compelled Bhattacharya to examine Aadhaar and India Stack much more closely—and to discuss the possibilities at length with Nandan Nilekani. It was soon evident to her that when completely implemented, the fee-based income banks now place a premium on would vanish. In turn, it would severely dent a bank's profits. If SBI was to stay relevant in the long run, the business model must be re-invented.

Bhattacharya has the credit of being the first head of a commercial entity that saw potential in deploying India Stack as a platform that could be harvested for profits. Until then, it (Aadhaar, primarily) was predominantly used by the government agencies hoping to fix leakages.

The linking of Aadhaar to subsidies made Bhattacharya realize that while Aadhaar can be an enabler, it can create problems as well. With their customer base increasing exponentially, banks found it difficult to service its customers—and complaints began to mount regarding waiting times.

SBI's technology team, though, had observed people's propensity to use cell phones. They thought up an app called NoQueue. When activated on an individual's handset, it told them of all the services that could be availed within a 5-km radius.

Once they chose the service they needed, the app would tell them how long it would take a branch to process their request. It would also let the customer know when the bank was ready to process their request. Wait times and complaints declined.

Multiple technology solutions started to emerge from the bank basis its experience in dealing with India and deploying various components of India Stack.

The top team at SBI was happy to allow all of this experimentation to go on. Bhattacharya was reasonably clear it was only a matter of time before an entity like Reliance Jio would attempt to encroach into banking and that to protect SBI, India Stack must be deployed tactically.

Bhattacharya had exchanged notes on this with peers from other parts of the world—like Francisco Gonzalez, executive chairman of BBVA, the second largest bank in Spain. 'Is it time to think about the business of banking as incidental to a banker?' they had asked of each other.

They were both on the same page—that it is indeed inevitable, and that they will have to alter their business models fundamentally; but what form their entities may morph into was still unclear.

When we asked Mrutunjay Mahapatra, the chief information officer at SBI, what his bank's future looks like, he didn't take much time to get to the point: 'Banking now is incidental to the SBI. This is now a technology company powered by data.' Had he said this even in late 2015, it may have sounded heretical.

What Mahapatra means is that SBI has accepted that banking in its current avatar may not earn it the kind of money it did in the past. In trying to understand the patterns embedded in the data generated by the bank, it has been pushing people to develop products and tools of all kinds to automate and solve problems. These can be sold to other entities that don't have the technical expertise to build it yet. To that extent, it can morph into a technology vendor as well and make up for losses from its traditional income sources.

The current regulatory regime, however, does not permit SBI to do that yet. There is much optimism, though, that it is only a matter of time before the RBI allows SBI to do so.

The optimism finds reflection in that SBI has set aside a fund of Rs 2 billion to assess start-ups in the fintech ecosystem. This is intended to scout for start-ups it can partner with to build products. When the plan was presented to the board, it agreed right away. The submission made to them was that with the emergence of technology, there are multiple challenges emerging. Some of these are best addressed by engaging with smaller and nimbler companies.

The Pioneer's Stake

Kishore Biyani of the Future Group does not come across as a data junkie. Instead, he seems more like an anthropologist—the kind of man who likes to spend time observing people in different retail environments. He never appears to be in a hurry.

Something changed in April 2017, though. A man who had no problems with long pauses had morphed into a man in a hurry. Whatever had happened?

Biyani thought he could spot the sliver of an opportunity in Rajasthan. After examining its finances, the state government thought the public distribution stores or '*ration ka dukaan*' as they are more popularly called, must be revamped. Private participants were invited. The Future Group articulated interest and the government called Biyani and his senior team to consult.

After studying the stock in these stores, the team at Future Group got back to the government and said that in its current avatar, participation was not a feasible proposition for them. On average, each store stocks 10–12 discounted items. This is not good enough to meet any household's requirements. To make a store viable, their experience suggested, at least 300 items must be made available. The state government thought that this may be difficult because stores that offer subsidized products are prone to pilferage by unscrupulous elements.

One way to weed these elements out was to build a system that can identify those who are eligible for it. The government suggested a caveat: give preference to women and senior citizens.

The plan was given a go-ahead. The team at Future Group got down to work on designing the stores, what items to stock, etc., and created a brand called Annapurna Bhandar. As things are, in Rajasthan, a little over 6,000 of these stores are operational.

While it has allowed the state to target those who need subsidies, on its part, Future Group has managed to gain an entry into parts of India that were otherwise inaccessible. It is gaining insights into how customers in this segment live and what they prefer. For instance, that people in rural Rajasthan like tomato ketchup is a learning that would otherwise not have occurred to the team.

As things are, with just a little over Rs 1 billion in turnover from this system, it isn't much to talk about in Future Group's overall scheme of things. In the longer term though, there is tacit acknowledgement that potential exists to make the business grow larger.

The learnings that are emerging are being ploughed back into Future Group's systems, and plans are being put into place to diversify into other states. Andhra Pradesh is among those that have called for a bid and active conversations are on. Governments from the North-east, Karnataka, Uttar Pradesh, and Madhya Pradesh are watching this play out with much interest as well.

The Global Take

Microsoft has a problem. In the developed world where it was the most dominant entity, it has been outgunned by Facebook, Amazon, Netflix, and Google, with its offerings seen as being expensive. To get Microsoft back to its pre-eminent position is one of CEO Satya Nadella's stated objectives.

Nadella has placed Microsoft's interest to work with India Stack on the record[10] as well. 'In our case, we would love our services such as Office 365 to be great participants in India Stack. I should be able to login using Aadhaar, be able to use any of the applications using the identity system that every Indian uses. Also, any entrepreneur who wants to build for India Stack should be able get the core infrastructure through our cloud infrastructure.'

That's probably why he suggested to Nandan Nilekani, Shankar Maruwada, Sanjay Jain, and Pramod Varma that they spend time with him in Bengaluru when he visited India last in August 2017. When the request from his office for a meeting came though, these gentlemen had no idea what Nadella had in mind.

Once they got together, Nadella politely asked if he may make a presentation. He then took to the whiteboard for 45 minutes and articulated his understanding of Aadhaar and India Stack. Basis that, he articulated the future he imagined for Microsoft if it adopted India Stack.

Those who work with Nadella suggest that he may have argued thus: 'Microsoft is an entity that caters to a developed world and is home to 1 billion people. The developing world though has 6 billion people with an average annual median income in the region of $1,500 per annum. If we can work off India Stack, Microsoft can increase its addressable market size. What may your thoughts be?'

Nadella was thinking along much the same lines as Mukesh Ambani. Is it possible to make sachets out of software? Netflix has already piloted the idea in India. Between late 2017 and early 2018, the entity has started commissioning original content tailored to meet the demands of Indian audiences.[11] It has even budgeted $8 billion to commission original 'evergreen' content such as drama, comedy, documentaries, and movies for 2018. While these programmes certainly get the attention of audiences when released, they retain their significance over time as well and earn money as more people discover them.

One reason why it did not make sense in the past to create evergreen content and charge small amounts of money was that transaction fees were high. But as the cost of transferring smaller amounts of money reduces, it becomes viable to build content for the long term. A company can then wait it out and hope to attract wide audiences in the longer run. Those who succeed will propel the Internet forward as a marketplace of ideas, experiences, and products—a marketplace of content.

If this is the future, a monthly or annual subscription may be out of the average Indian's budget. However, if they are allowed the option to pay a few rupees each time they want to watch a movie or an episode from a television show, India Stack, with UPI built into it, is a viable proposition. That is why many digital players see long-term merit in India.

Nadella was wondering if the same rationale can be extrapolated to Microsoft's software business. Much like with digital content, the incremental cost of creating another piece of software is zero.

With UPI in place, Microsoft can see multiple options. Potentially, new systems can be implemented that allow users to pay each time they may want to use a piece of software. For instance, somebody may want to use a licensed version of Microsoft's spreadsheet Excel just once. The document can be stored on Microsoft's Cloud so it is never lost. People may not be able to pay a monthly fee, but could pay small amounts on a pay-as-you-go basis.

If this model can be made to work, it can potentially be exported as a solution by Microsoft to other parts of the developing world as well where people face the same problem.

What emerged out of these deliberations is still a mystery. What we do know, however, is that businesses across the world are in an uncomfortable place. It's no surprise then, that to most people, data-driven businesses such as Google, Facebook,

Amazon, and Apple are the way to go. Yet, Scott Galloway, of the NYU Stern School of Business, describes these Internet giants as 'The Four Horsemen' in his best-selling book, *The Four*. He also makes some more predictions. 'The only thing I am comfortable saying is they will all go out of business, all disappear within 50 years.'[12]

The Four also speculates on who the next horseman might be; Microsoft was not among the companies that may go out of business.

The Leader

That banks would be unbundled in the future was clear to Bill Gates as early as 1994, when he famously said, 'Banking is necessary, banks are not.'[13]

Aditya Puri, managing director of HDFC Bank, is said to have been mighty disturbed when he read the report Morgan Stanley published on 26 September 2017. While the authors of the paper, titled 'India's Digital Leap—The MultiTrillion Dollar Opportunity', said that they remain optimistic about the Bank's outlook for the next decade, three lines on page 84 of the report on where they may be wrong got his attention:

a) greater than expected competition in retail on pricing;
b) slower than expected CASA [Current Account, Savings Account] growth;
c) higher than expected competition in processing and other fees.

Stripped of all jargon, they were saying HDFC Bank's future was under threat. Basis his own assessment, stock prices, and inputs from multiple sources, everything suggested all is well. But this report was insinuating that they were all mistaken.

The analysts could see the writing on the wall. If service charges were taken out of HDFC Bank's balance sheet, there was much to lose: In 2017, 27% of HDFC Bank's profits[14] emerged out of service charges of all kinds. The A-team was called in. Some intense brainstorming later, those who were present recall that he said he could hear the sound of 'cognitive dissonance'.

It was clear, Puri told those who were assembled there, HDFC Bank would have to reinvent itself. Before he had called his senior team in, Puri had had multiple one-on-ones with the people at India Stack and others who had studied it closely. When Puri connected all the dots, he wasn't pleased with the final picture: Only those at HDFC Bank thought that it was digitally ready.

A team from iSPIRT was called in to present personally to him. He asked them to be brutal and not spare any punches. What the bank lacked, they told Puri, was a technology backbone; their digital services were 'just a wrapping'. This, at a time when entrepreneurs from other industries were already experimenting with India Stack. What the entrepreneurs see, explained the iSPIRT team, is a set of tools that they can use to get into domains they may otherwise not have access to—banking included. It is inevitable that these entrepreneurs get into markets Puri's team considered unviable at the moment. If he needed any evidence for iSPIRT's claims that entities like HDFC Bank would come under scrutiny in the long run, the Morgan Stanley report had subtly driven home the point.

Equity analysts based out of India are bullish on the stock and are unanimous HDFC Bank looks impregnable. HDFC

Bank did not respond to questions around its strategy for India Stack. Sources have it though that Puri sent a trusted team to Bengaluru towards the end of 2017 to examine the architecture of India Stack and report back to him if there is merit in embracing it.

The Verdict

When we asked Bindu Ananth of the Dvara Trust if she can spot a thread that binds all of these models clearly, she thinks the future is fuzzy. But then, from her perspective, the process of building new business models has only begun.

In banking, for instance, even lending, Ananth says, can be unbundled and stripped down into more consumer-friendly products. A student's loan needs may be very different from those of a salaried professional in an urban city, or an entrepreneur in a village. Basis the research the Trust has conducted, Ananth says, India cannot be looked at as one country, or three, or even seven as some studies have suggested, but a federation in excess of 700 countries—because there are more than 700 districts in India. And each district has populations that are as large and distinct as countries in the world are. Each of these 'countries' need distinct entities to serve their own unique needs.

The world has never dealt with something like this before. From a policymaker's perspective, what India needs desperately now is a mechanism to begin grappling with all of what it is staring at. A regulatory body headed by someone like a chief data officer is a good place to start. Here, too, there are no formal voices yet—only informal conversations have begun.

Providing a uniform identity like Aadhaar is only the starting point that binds these 'countries'. Work can now begin on multiple fronts to aggregate the data, make sense of it all, and innovate relentlessly.

This also opens the doors for new players to emerge out of nowhere and think up something totally unheard of. To crown anyone a winner, whether they be the hopefuls or the hot favourites, would be too early—and rather naive.

Notes

1. PTI, 'Aadhaar Covers 89% Population', *Times of India*, 7 March 2018, https://timesofindia.indiatimes.com/business/india-business/aadhaar-covers-over-89-population-alphons/articleshow/63202223.cms, viewed on 10 August 2018.
2. Dan Frommer, 'Watch Mary Meeker Give Her 2017 Internet Trends Report', *Recode*, 2 June 2017, https://www.recode.net/2017/6/2/15731874/watch-mary-meeker-2017-internet-trends-report-full-video-code-conference, viewed on 9 July 2018.
3. Puja Mehra, 'All You Need to Know about Payment Banks', *The Hindu*, 20 August 2015, https://www.thehindu.com/business/all-you-need-to-know-about-payment-banks/article7561353.ece, viewed on 14 July 2018.
4. *Mint*, 'RBI Tells Paytm Payments Bank to Stop Adding New Customers with Immediate Effect', 10 August 2018, https://www.livemint.com/Companies/IEa620FHKxyja4b36qWr0J/RBI-tells-Paytm-Payments-Bank-to-stop-adding-new-customers-w.html, viewed on 11 August 2018.
5. Aparna Iyer, 'Paytm Faces Red Queen Test to Make Valuation Stick', *Mint*, 26 January 2018, https://www.livemint.com/Money/hq6PY8ARcX9tNBlrhf2feK/Paytm-faces-Red-Queen-test-to-make-its-valuation-stick.html, viewed on 9 July 2018.

6. Crunchbase, 'One97 Communications', https://www.crunchbase.com/organization/one97-communications.
7. Nadeem Unuth, 'How Many Megabytes for One Minute of Conversation?', *LifeWire*, 13 May 2018, https://www.lifewire.com/megabytes-for-one-minute-conversations-3426705, viewed on 9 July 2018.
8. *Indian Express*, 'Reliance Jio 4G Launch', 1 September 2016, http://indianexpress.com/article/technology/mobile-tabs/reliance-jio-4g-launch-ril-agm-live-3007424/, viewed on 9 July 2018.
9. Paranjoy Guha Thakurta and Aditi Roy Ghatak, 'The Immaculate Conception of Reliance Jio', 4 March 2016, http://paranjoy.in/article/the-immaculate-conception-of-reliance-jio, viewed on 9 July 2018.
10. Shweta Modgil, 'Satya Nadella Thinks Highly of India Stack', *Inc 42*, 20 February 2017, https://inc42.com/buzz/satya-nadella-flipkart-microsoft/, viewed on 9 July 2018.
11. Shweta Modgil, 'Netflix Sees a Potential of Adding the Next 100 Million Customers from India', *Inc 42*, 26 February 2018, https://inc42.com/buzz/netflix-india-reed-hastings/, viewed on 9 July 2018.
12. Therese Poletti, 'Amazon, Apple, Google and Facebook will All Go Away within 50 years, Says Author', *MarketWatch*, 19 October 2017, https://www.marketwatch.com/story/amazon-apple-google-and-facebook-will-all-go-away-within-50-years-says-author-2017-10-17, viewed on 9 July 2018.
13. Markus Filkorn, 'Banking Is Necessary, Banks Are Not', Capgemini Consulting, 5 July 2016, https://www.capgemini.com/consulting/2016/07/banking-is-necessary-banks-are-not-how-banks-can-survive-in-the/, viewed on 9 July 2018.
14. HDFC Bank Investor Presentation, 2018, https://www.hdfcbank.com/assets/pdf/Investor_Presentation.pdf, viewed on 9 July 2018.

6

The Naysayers

ON THURSDAY, 1 FEBRUARY 2018, JUSTICE D.Y. Chandrachud, one of the five judges hearing a case on the constitutionality of Aadhaar, was irked by the tone of the arguments made by lawyer Shyam Divan. A year earlier, Chandrachud had been one of the nine judges who had unanimously ruled that privacy was a fundamental right in India. He and his fellow judges were now hearing a case on whether Aadhaar violated that right, among others.

Divan, who had developed a fan following among Aadhaar critics for his passionate arguments against the identity programme, was discussing the written pleadings made by his clients, who wanted Aadhaar to be stopped. The judge and the lawyer disagreed on the scope of the pleadings. The argument got intense.

The judges had been hearing Divan's arguments for a few days. It was clear that he was passionate about the subject. But he was also pushing it a little too hard.

At one point during the argument, Justice Chandrachud said, 'There is no point raising your voice every time. Constitutional issues cannot be argued on hyperbole. Should we stop asking questions?

And why do you keep telling us that we will be known as "Aadhaar judges"? What is the way of saying unless you agree to our view, we will say you care less about the community? We are not answerable to anyone here as a judge except to our conscience.'[1]

He went on, 'The moment we ask questions, we are attacked as if we are committed to an ideology or anyone. If that is so, I plead guilty to the charge. We are not here defending the government, nor are we going to follow any NGO line. What is this way of saying that either you are with me or you will be branded as "Aadhaar judge"?'[2]

Divan realized that he might have overstepped, and apologized.

Such intensity is not entirely uncommon in the Supreme Court. Both the nature of the case and the circumstances under which it was being argued were fraught with tension. After all, Aadhaar touched practically everyone in the country, and had become an obsession for some of the most powerful voices in civil society. The government itself was betting on Aadhaar for many of its flagship social programmes. Cases on Aadhaar were being fought in the courts since 2012, under intense media gaze and all-round interest. Over time, cases have piled up, one on top of the other—all questioning the methods or validity of Aadhaar. In January 2018, the court started hearing what is probably the most important case on Aadhaar, bundling together about 30 separate petitions.

Apart from the high stakes involved, the Supreme Court itself was going through a crisis. Weeks before the case started, four of the most senior judges addressed journalists in an unprecedented press conference to complain about the chief justice on the allocation of cases. They weren't talking about Aadhaar. But uneasiness hung heavily over the judiciary—and the tension spilled over everything.

Such heated exchanges were not limited to officialdom alone. Last year, an acquaintance got into a Twitter argument on Aadhaar—and it stretched, off and on, for three days. One night, while lying in bed, staring at the dim reflection of the streetlight on the blades of the ceiling fan, it occurred to him that instead of sleeping, he was thinking about that debate. In fact, he hadn't slept well the previous two nights either.

What began as a gentle disagreement about some point on Aadhaar had descended into nasty attacks. His attempts to be reasonable were met with ridicule. His intelligence and integrity were questioned. Some of his tweets were twisted and he felt people had ganged up against him.

And these were not anonymous trolls. Some of them were acquaintances, people he had met, people he knew professionally. That night, he promised himself, that he would stop engaging, that he'd forget the whole episode, and focus on his work and his family. 'It got too personal,' he told us a few days later. There was a tinge of sadness in his voice.

Over time, the debates have become more intense, the attacks more personal, and Hitler's name has been mentioned much more often—quite in line with Godwin's Law, which says that as an online discussion grows longer, the probability of a comparison involving Hitler gets higher. The intensity gets reflected in the news: 'Getting Aadhaar is very much like getting AIDS. The government can no longer distinguish citizens from residents. It cannot distinguish legal residents from illegal residents,' a column in the *Economic Times* said.[3] A business news editor of its television channel, ET NOW, got trolled because the channel interviewed the UIDAI CEO.

Among those passionate about Aadhaar, the issue is not one of technology, but one of morality. Naysayers accuse its supporters of turning a blind eye to Aadhaar's exclusion of the poor, the risks it poses to privacy, and its vulnerability to misuse. Its proponents feel that the anti-Aadhaar brigade defends a status quo that has kept millions in poverty, that it obsessively focuses on Aadhaar's flaws, which are fixable, and fails to set the risks to privacy and fraud in the context of the broader digital world. Both parties suspect the integrity of their opponents.

What Justice Chandrachud felt—'the moment we ask questions, we are attacked as if we are committed to an ideology or anyone'—was just a reflection of what people who get into any discussion about Aadhaar feel every day. As former Planning Commission member Arun Maira puts it, 'There is a feeling that you are either with us or against us.'

Why should civil society—which presumably has public interest in mind—want to junk a programme that the government asserts was created for inclusion, better delivery of public goods, and for innovation that will help society at large?

The answer to these questions lies in the stories of social sector activists and technologists, and how they came together to fight what they believe is one of the biggest wars of our times.

Identity or Identification?

'When I first read about Nandan becoming the chairman of UIDAI and his vision to give an identity for every Indian, I felt happy. At that time I believed it will empower people,' Usha Ramanathan, one of the most vocal and earliest critics of the project told us. 'I had seen how providing an identity to the poor could help them. It happened with the voter ID cards. So, in the beginning I was all for it.'

'Later, UIDAI had organized an event in Bengaluru. They invited academics, researchers, and activists to explain the project. Nandan Nilekani gave us a presentation on their vision for UID, and listening to that, it became clear to me that it was not an identity project. It was an identification project. It was right there in the name, Unique Identification Authority of India.'

'An identity project would give power to the people, it will empower the poor by giving them an identity. An identification project empowers the state. That's a big difference,' she said.

We met Ramanathan at India International Centre in New Delhi one early morning in June 2017. She introduced herself as a 'Khan Market Liberal', a lighthearted reference to Nilekani's description of the critics a couple of months earlier. At the cafe inside, she suggested we sit away from the other members to ensure our recorder doesn't inadvertently catch their conversations.

Ramanathan is one of the most prominent voices in legal research. After studying law at Madras University, the University of Nagpur, and Delhi University, she immersed herself in the intersection of law and society. Earlier, among other things, she had worked on security and welfare laws to improve the lives of beggars. She had also lent her support, expertise, and voice to some of the defining cases of civil society, including the Bhopal gas disaster, the campaign against Narmada valley dams, and slum eviction in Delhi.

For a long time, she used to go around on a scooter. She still doesn't carry a mobile phone. It's not that she doesn't like technology. As she once explained, 'For me, when it comes to any technology, the question is whether I will be using it or whether the technology will end up controlling me.'

Her concerns about Aadhaar stem from the same question. Over the years since her first insight about the project—that it is not an identity project, but an identification project—she found that the actions of the government and UIDAI had only justified her concerns. 'It was sold to us as if it was voluntary, but now, it has in effect become mandatory. It was sold as a cure for all the problems facing the country, but the government hasn't honestly answered questions about its impact,' she told us.

The Jholawala Economist and His Friends

Jean Drèze, who now teaches economics at the University of Ranchi and has co-authored books with Nobel prize winning economist Amartya Sen, has been a vocal critic of the project right from the word go. To slot him as an academic who studies development economics would be unjust to what he believes in and what he does.

The son of Belgian economist Jacques Drèze (whose contributions include a classic and one of the most widely cited papers on equilibrium), Jean did his PhD in India and eventually took Indian citizenship. Even back in 1979, he lived in a slum in Delhi to better understand the people he was researching on (perhaps making him the poster boy for the diehard do-good, jhola-sporting activist). It's a method he persists with till today, and that has ingrained in him a deep respect for the poor.

After 15 years of research on hunger and famines, one is perhaps entitled to feel like an 'expert' of sorts on these matters, he wrote in the *Economic and Political Weekly*. 'Yet I did not always find myself better equipped than others to understand the practical issues that arose in this situation.

At times, I even felt embarrassingly ignorant compared with local people who had little formal education but a sharp understanding of the real world. Some of them were curious about my collaborative work with Amartya Sen (who had become a household name in India after winning the Nobel Prize for Economics), but when I tried to explain to them the main insights of this work, they were not exactly impressed. It is not that they disagreed, but they just thought that the basic message was fairly obvious.'[4]

He doesn't stop himself from showing his connection to the real world. Once, in 2009, during an interview at NDTV studios, he brought a food basket, and while making his argument, pulled out an overripe banana, an egg, and a packet of milk, and spelt out the cost of each to underscore his point that getting nutritious food is expensive for the poor, and even for the middle class.

His was also an influential voice during a good part of the UPA regime under Prime Minister Manmohan Singh. The UPA was dependent on the Communist Party during its first term, and the common minimum programme leaned heavily on development and inclusion. Drèze was a member of the National Advisory Council, and he practically wrote the government's employment guarantee law, which assured at least 100 days of employment for the poor. What Drèze brough to the table was research, action, and influence.

His concern about Aadhaar was that it was trying to fundamentally disrupt the social security system that was emerging, not by itself, or by the benevolence of the government, but by back-breaking activism on the ground by generations of social workers. Going through his early criticism on Aadhaar suggests that it was twofold. One was that a disruptive initiative such as Aadhaar could lay waste the years of work building networks, legislation, and institutions—and replace it with a cash transfer programme. Two, it was in effect laying out a red carpet for corporate interest.

A 2010 piece he wrote captured his concerns.[5]

> The real game plan, for social policy, seems to be a massive transition to 'conditional cash transfers' (CCTs). There is more than a hint of this 'revolutionary' plan in Nandan Nilekani's book, *Imagining India*. Since then, CCTs have become the rage in policy circles. A recent Planning Commission document argues that successful CCTs require 'a biometric identification system', now made possible by 'the initiation of a Unique Identification System (UID) for the entire population ...' The same document recommends a string of mega CCTs, including cash transfers to replace the Public Distribution System.
>
> If the backroom boys have their way, India's public services as we know them will soon be history, and every citizen will just have a Smart Card—food stamps, health insurance, school vouchers, conditional maternity entitlements and all that rolled into one. This approach may or may not work (that is incidental), but business at least will prosper. As the *Wall Street Journal* says about the Rashtriya Swasthya Bima Yojana (which is a pioneering CCT project, for health insurance), 'the plan presents a way for insurance companies to market themselves and develop brand awareness'.

Another equally powerful voice in the National Advisory Council was that of Aruna Roy. Roy is a former IAS officer and co-founder of the Mazdoor Kisan Shakti Sangathan (MKSS) along with Nikhil Dey. Roy, Dey, and their colleagues kick-started the Right to Information (RTI) and employment movements in India. And their concern about Aadhaar was similar to Drèze's—far from empowering people it could become the most powerful tool to disempower people.

On a visit to the National Institute of Advanced Studies in Bengaluru in 2010, Dey gave a sense of how he looked at MGNREGA (Mahatma Gandhi National Rural Employment Guarantee Act) and RTI, which incidentally throws light on the reasons for his discomfort with a centralized, digital initiative like Aadhaar. MGNREGA and RTI, he said, represented a switch in power relationships in the society. MGNREGA allows every rural Indian to demand work, and they would get it in 15 days. It's a move from a legal system where the government was dishing out things, to a system in which people can demand entitlement and government is accountable for it.

Once that's in place, people will figure out a way to audit and account their entitlements. One common refrain among many who have worked with the poor is the way they used walls as a tool. Every detail is painted on the wall for everyone to see. (In Rajasthan for example, MKSS got the details of MGNREGA workers and the amount they received painted on a wall, reducing the chances of the agents siphoning off a part of wages. Transparency makes feedback loops work stronger. (If you recall, in Nigeria, the simple act of publishing in newspapers the amount spent on schools resulted in less leakage.)

To Usha Ramanathan, Jean Drèze, Nikhil Dey, and others, the big apprehension was that Aadhaar did not switch the power from the state to the people, like MGNREGA or RTI did, but it gave more power to the government. Those who ran the government, the bureaucrats and the politicians, held the digital switch that could provide or deny entitlements to the poor.

Those who were worried about the risks of Aadhaar were concerned about these twin issues: that the government would push Aadhaar in various schemes to citizens, and at the same time, create infrastructure compelling businesses to try out Aadhaar and related technologies. They were marketed as if they would empower the citizens and consumers. But the result could be quite the opposite.

But these early opponents of Aadhaar would get support from a least expected quarter—from Bengaluru, metaphorically speaking.

Seeds of Exclusion

Reetika Khera is an economist very much in the mould of Jean Drèze, in that her research involves feet on the ground. Like many activist-academics, she puts her research ahead of personal comforts. One evening, she was at Amnesty International's offices in Bengaluru. It was 4:45 pm, and she still hadn't had her lunch. Someone managed to pack her something from a cafe nearby. Peeling her eyes off the computer screen long enough to thank her profusely, Khera had her 'lunch', eyes back on the screen, working the keyboard with one hand.

In 2013, when A. Babu was the collector of East Godavari district in Andhra Pradesh, Khera went there to study a pilot project involving the use of Aadhaar in the PDS. Babu was piloting an end-to-end computerization of PDS at the time. Instead of having to present a paper document or a smartcard at the shop, users could authenticate themselves by providing their Aadhaar numbers and biometrics. If the fingerprints didn't work, they could use mobile OTPs. It took about a year of ground work to launch the pilot.

During her visit Khera found some things that she liked—the computerization of the back end, and the ePOS terminals. However, she also found many of the problems that would haunt the other projects in other cities. People were not too excited about the change—the earlier method took less time and was not dependent on technology. Some of the Aadhaar numbers were not linked to the existing database. One lady missed out on ration because she could not authenticate herself. The ghosts that were found could have been found through a door-to-door survey rather than biometric deduplication. In short, she was fine with computerization, but not with Aadhaar.

Aadhaar supporters argue that technology will improve, that designers learn from past mistakes and things tend to get better. Asked if she found things improving on the ground, Khera replied in the negative, pointing out to exclusions in Rajasthan, Jharkhand, and other places. (In September 2017, 11-year-old Santoshi Kumari died of hunger because she was denied ration as her Aadhaar was not linked to the system.)

However, towards the end of 2013, the same year that Khera studied the East Godavari pilot, it seemed as if the civil society had won a resounding victory over government imposition of Aadhaar. The UPA government could not pass a law to back Aadhaar. A standing parliamentary committee report submitted by former Finance minister and BJP leader Yashwant Sinha adequately reflected their views. The government had put a stop to the LPG DBT scheme (though it was not because of the efforts by Drèze and fellow activists, but because of lobbying by Kerala LPG distributors). And BJP, which seemed to be gaining ground every passing day, driven by the charisma of Narendra Modi, was roasting the government in its election campaign on account of Aadhaar. Nilekani, who stood for elections on a Congress seat from Bangalore South constituency, lost to his BJP rival.

Second Innings

The mood among those who built Aadhaar was bleak. Five years of their work might be wiped off in one go. However, in the first few weeks and months of Modi rule, it was as if a one-day cricket match had been suddenly converted into a test match. Aadhaar got to play a second innings. Now, Modi stepped on the gas. Enrolment had gained speed, and bureaucrats started to think of ways in which they could use Aadhaar.

Meanwhile, the team that built Aadhaar was busy building what was eventually presented as India Stack. UIDAI, for all practical purposes, had turned into a government department with little of the creative energy that marked its early days.

One source of that energy currently was Sharad Sharma, a technology manager, investor, and evangelist for technology products in India. He parted ways from IT industry body Nasscom to set up iSPIRT, a think tank with a missionary zeal, with his friends to transform India into both a producer and consumer of cutting-edge technology. The best way to do that, his reasoning went, was to build digital infrastructure and let entrepreneurs build products and solutions specifically designed for India (instead of copy-pasting products and business models from the US or China). Aadhaar was one such digital infrastructure—and the group of volunteers under the iSPIRT umbrella wanted to build the remaining pieces.

Together, they would enable what Nilekani calls presence-less, paperless, and cashless transactions. This in turn would bring down transaction costs, setting off a virtuous cycle.

Facebook and the Rise of Tech Activists

Towards the end of his 10-page document entitled 'Is Connectivity a Human Right?',[6] Facebook boss Mark Zuckerberg detailed how his company's initiative, internet.org, a platform that would allow the poor to access certain specific sites without having to pay for it, would be good for different groups:

> This is good for people because they'll have an affordable way and a reason to connect to the internet and join the global knowledge economy.
>
> This is good for mobile operators because they'll have more customers who want to buy more data, which will increase their profits and help them invest in building out the networks.
>
> This is good for phone manufacturers and technology providers because more people will want better devices, which will push the industry forward.
>
> This is good for internet services because the efficiencies we'll all drive will make it easier and cheaper for the next 5 billion people to access their services.
>
> This is good for the world because everyone will benefit from the increased knowledge, experience and progress we make from having everyone connected to the internet.

Zuckerberg wrote this memo in 2013, but it was not until 2015 that Facebook launched internet.org in India. His mission didn't go down well with a group of technologists and technology policy activists. Expecting nothing good to come of it, they were all set to fight it.

The most prominent among these was Nikhil Pahwa, who cut his teeth in online media running ContentSutra, a part of entrepreneur Rafat Ali's ContentNext media, and later launched Medianama, a website that was focusing on mobile and digital news. Soon after Facebook announced its internet.org plans for India, Pahwa saw that it would violate the principle of net neutrality. Net neutrality dictates that telecom companies treat all data equally. However, by zero-rating some websites—in this case, the sites that would sign up with Facebook's platform—telecom companies would end up violating the principle.

An event organized by Pahwa's Medianama in Bengaluru on privacy gave a sense of how he likes to work. Most events tend to have people on stage speaking most of the time, with very little time allotted for questions (and often, only questions). At this event, however, the audience were encouraged not just to ask questions but also share their views. Pahwa went around the room, pulling people out to dwell on a topic of their interest or expertise. It bridged the gap towards an unconference, where participants drive the discussions, avoiding the constraining 'structuredness' of conventional events.

In the online campaign for net neutrality Pahwa and his fellow volunteers nudged citizens to send mails to the telecom regulator. All India Bakchod, a comedy group, produced a short movie on net neutrality, garnering more than 3.5 million views. And people responded by sending thousands of mails to TRAI. Facebook responded by encouraging its own users to send messages in support of internet.org (it had changed its name to Free Basics by then). According to one report, Facebook spent $30 million on advertising.

Ultimately, the decision fell to the regulator—TRAI chairman R.S. Sharma. There was tremendous pressure. While the debates were at their peak, Narendra Modi visited Facebook headquarters. Zuckerberg, in turn, not only visited India, but also personally spoke to a number of Internet entrepreneurs, trying to convince them of the merits of Free Basics. TRAI ruled upholding net neutrality. In February 2016, Free Basics was banned in India. And through the fight tech activists realized that, if they could get together, they could influence policy.

Getting 'Panned'

By April 2016, Aadhaar had crossed a billion enrolments. UIDAI announced that 93% of adults, 67% of children aged 5–18 years, and 20% of children below 5 years of age had an Aadhaar. It claimed that linking Aadhaar with various schemes was showing benefits in terms of cost savings. The government, it said, saved an estimated Rs 146.72 billion through DBT in cooking gas, Rs 23.46 billion in PDS across Andhra Pradesh, Telangana, Puducherry, and Delhi, Rs 2.76 billion in scholarship amounts across Andhra Pradesh, Telangana, and Punjab, and Rs 660 million in pensions across Jharkhand, Chandigarh, and Puducherry.

It also announced that 254.8 million bank accounts were linked with Aadhaar. Over 71% LPG connections, 45% of ration cards, and 60% of MGNREGA cards were also linked with the unique ID. UIDAI was doing 4 million authentications a day. And it had done 84 million eKYC transactions. In short, Aadhaar enrolments had scaled up, UIDAI had demonstrated its capacity to handle high volumes of transactions, and its use in various schemes was showing benefits.

Over the next months and quarters, the government would mandatorily link more services to Aadhaar, including PAN cards, bank accounts, and telecom services—none of which were subsidies. Ramesh Srivats, an entrepreneur known for his witty commentary on Twitter, made this observation. 'Got it. It is compulsorily mandatory to voluntarily get yourself an Aadhaar card.'

For the technology activists, though, the idea of linking Aadhaar to everything was not just about whether it was mandatory or voluntary. Pretty much like Zuckerberg's memo, the government was telling each interest group how Aadhaar was good for it and for the country. And just as internet.org was violating net neutrality, Aadhaar was violating or could violate privacy. And just as how it fell upon TRAI to make the final decision, it was up to a few judges in the Supreme Court to decide on the fate of Aadhaar. Convince them, and the job is done.

However, net neutrality and privacy are two different animals. Net neutrality has a specific meaning; privacy is a broader concept. The question on privacy is central to matters as varied as laws on homosexuality, clinical trials, police action, and the practices of big tech companies—not just Aadhaar. The Indian government was already running a number of programmes whose stated purpose is surveillance, with serious privacy implications. More importantly, consumers were increasingly sharing more personal data—though perhaps without realizing exactly how it would be used—with tech companies, even as the capacity of tech companies to crunch that data is growing exponentially.

In fact, the privacy argument was not the discovery of the technology folks. Those who opposed Aadhaar because it could lead to exclusion had, in fact, used the privacy argument in court to say that it was wrong for the government to link subsidy schemes to biometric ID, because it violated the privacy of beneficiaries. In 2015, Mukul Rohatgi, the country's attorney general, arguing in the Supreme Court, countered it by saying, 'Right to privacy is not a fundamental right under our Constitution. It flows from one right to another right. Constitution makers did not intend to make Right to Privacy a fundamental right.'[7]

The matter however was not simple, because there were two contradictory judgments on the issue by the Supreme Court. One bench of five judges had said that privacy was a fundamental right, but an earlier judgment that said that it was not had seven judges on the bench. As a result, the Supreme Court said it would form an even larger bench to decide on the issue. Meanwhile the cases that came to the Supreme Court should be argued on other issues.

One such case that came up was on whether it was legal for the government to make linking of Aadhaar with PAN card mandatory.

Aadhaar 'Disclosures'

Srinivas Kodali has a BTech from IIT Madras and is passionate about intelligent transportation systems and open data. Soft-spoken and polite, his demeanour can hide the intensity with which he pursues his passion. Once, while presenting at an event in Bengaluru with Sam Pitroda, who helped kick-start the telecom revolution in the country, and Carl Malamud, who runs Public.Resource.Org, an NGO that has placed hundreds of millions of pages of government information online, he said public information belongs to us, it's not government property. Kodali and his fellow volunteers had just archived nearly 150,000 gazette notifications from the government of India, making it accessible and searchable.

Kodali was passionate about open data, but he also knew that doing it wrong could end up disempowering people, instead of empowering them. When he looked at Aadhaar, he was not happy with what he found. 'Many who criticize Aadhaar on the grounds of technology come from an open source background,' Kodali told us. 'But the bigger issue, whether you believe in open source or not, is around security, and how safe your data is. Government departments don't follow good technology practices. By letting them handle their personal data, the public is opening itself to security risks. The system is leaky, and it's leaking data,' he said.

In May 2017, just as the Supreme Court was starting to hear the case about linking PAN with Aadhaar, Kodali along with a researcher from the Centre for Internet and Society released a report titled Information Security Practices of Aadhaar (or lack thereof): A documentation of public availability of Aadhaar numbers with sensitive personal financial information. It revealed that government websites had made 135 million Aadhaar numbers public.[8]

The scale—135 million—captured media attention, and the report received wide coverage. The hashtag #AadhaarLeaks started to trend. Later, the Centre for Internet and Society said it was not a leak, but a disclosure. The government departments were complying with Right to Information laws that mandates departments to reveal details of those who receive subsidies from the government. The transparency, it was believed, would help plug leakages.

A Misstep

Sitting in one corner of Bengaluru, Sharad Sharma did not look at the developments the same way. Those who have met and listened to Sharma use words such as 'logical', 'persuasive', 'evangelical', but most of all, 'combative' to describe him. He is knowledgeable and passionate about certain themes—India's potential to create software products, the need to expand the market to lower layers of the pyramid, and of course India Stack, which includes Aadhaar, eSign, eKYC, UPI, and DigiLocker. And when he talks it's as if his main purpose is not so much to share his knowledge as it is to somehow transfer his passion to the audience, even if it's an audience of one.

To Sharma, and many others who cared deeply about Aadhaar (and India Stack) the report was not just an attempt to give feedback to the government to fix data protection practices in its departments. While they were technically correct, the impression they created was that the core Aadhaar database was leaky. This generated fear, uncertainty, and doubt about Aadhaar. The Aadhaar evangelists did not see it as an isolated incident, but a concerted campaign against Aadhaar—aimed at influencing the judges who were hearing the cases in the Supreme Court. It was not entirely unfounded. A person who had a chance to review the conversations in an online group of activists and journalists opposed to Aadhaar said the group members discussed the headlines, hashtags, and social media promotion of articles that highlighted the risks and weaknesses of Aadhaar and Aadhaar-enabled applications. The name of the group captured their feelings about the project: Aadhaar Apocalypse. Besides the well-known critics of Aadhaar, there were also a number of anonymous accounts campaigning for the ban of UID. They were not sure what their agenda was. They could well be accounts operated by people who did not have the best interest of the country in mind.

Sharma did not want to just let it go. He wanted to respond and provide a fuller picture. What he did next would haunt him, and in some ways, undermine the work done by iSPIRT. At iSPIRT's Koramangala office, he assembled a few young engineers, created anonymous handles, and started responding to tweets on Aadhaar. At that time, the reasoning was that other Twitter users will look at their comments for what they were, without getting biased by their affiliations. But as every Twitter user knows, anonymity also takes away the responsibility, skin in the game, and soon some of them were trolling Aadhaar critics.

One such critic who got unfairly trolled was Kiran Jonnalagadda. Jonnalagadda runs an organization called HasGeek, which, by its own description, 'creates discussion spaces for geeks'. These events—focusing on topics ranging from javasript and android platform to data analysis and fintech (besides one devoted to fitness)—attract techies interested in networking, and enhances their knowledge. Jonnalagadda is a geek himself, and started capturing and analysing the tweets by these anonymous accounts—and it didn't take long to realize that one was operated by Sharad Sharma himself. (Sharma had on occasion switched from one of the accounts to his own during an argument. And had given his own phone number for second factor authentication).

Jonnalagadda published a piece in Medium listing the offensive tweets and outing Sharma. Sharma, who was on a trip to the US, apologized in a tweet—'On my flight back from the US, I reflected on my recent behaviour on Twitter.... I unreservedly apologize to all who were hurt ...'—attaching a longer note promising that it would never happen again.

The impact of the episode showed on iSPIRT. A number of technologists and entrepreneurs who had associated with the think tank understood why Sharma acted that way, but also felt disappointed that he did what he did. Since Sharma was closely identified with iSPIRT, the entire trolling episode was seen as officially sanctioned by UID. (A review of the tweets by the anonymous handles showed that less than 5% of them were abusive. A majority of them were civil exchanges focused on fact checking and calling out mistakes).

In fact, at that time, iSPIRT was discussing plans on how to engage with the critics. It had formed a team called Sudham to engage with them. In a typical think-tank style, they had divided the detractor universe into four quadrants based on levels of knowledge, and willingness to engage. For each of these segments they were planning a content strategy. Jonnalagadda had published a few slides from a deck, suggesting that trolling was in fact a part of its strategy. iSPIRT insists that the Sudham team had nothing to do with trolling. It was meant to share information regarding Aadhaar and India Stack, which could counter the negative propaganda on Aadhaar, and allay the fear, uncertainty, and doubt. However, the trolling incident had created such a backlash that they shut the project down.

Informed, Fearful, and Engaged

One name that figured among the detractors—under the quadrant 'Informed, Fearful, Engaged'—was Anand V. Anand had published a couple of well-argued pieces in Medianama (the site run by Nikhil Pahwa), and was looking at Aadhaar from his own vantage point.

Anand is a cybersecurity professional working for a technology company offering cloud services. Like Usha Ramanathan, he too rides a scooter to travel within the city and doesn't use a mobile phone for calls, even though he carries one to access services such as Ola. He is hungry intellectually, with deep interest in philosophy, mathematics, and, of course, technology. He says he only learnt about India Stack when he read a piece by Haresh Chawla in *Founding Fuel* and his interest got piqued. He started reading the extensive documentation that UIDAI had published on the Web (even though of late it has been pulling them off). Anand's mother used to work in Tamil Nadu civil supplies and was familiar with how things worked in the government, in the last mile, and in the bureaucracy.

The cybersecurity professional in him could see where the weak points were in the system as a whole. Whenever any news of a fraud or a breach broke he could connect the dots. Besides, as a white hat hacker, he looked for and found vulnerabilities in the system, duly reported them to UIDAI, gave it time to fix them, and wrote about it. He also started analysing data put out by the government agencies—pension-related data from Kerala, PDS-related data from Telangana—and started looking at how the schemes were performing on the ground. He could connect the dots between design decisions, implementation, impact. While most technologists were primarily looking at Aadhaar from a privacy perspective, he tried to take a holistic view, and saw how everything was connected.

Meanwhile, the Supreme Court heard the case on whether privacy was a fundamental right, and unanimously ruled that it was. For civil society, it was a victory. Aadhaar supporters welcomed it too. The judgment said that while the right to privacy was fundamental (which essentially means a citizen's right to privacy is protected against state action), it was not absolute. Government can restrict the right on the grounds of legality, need, and proportionality.

In a conversation with the Founding Fuel community on Slack, Rahul Matthan, a partner at Trilegal who focuses on technology practice, explained it thus. 'This is actually a pretty standard formulation against which fundamental rights are tested. So there are a number of precedents to guide us. Essentially there must first be a law—so no notification or government order can deny you your privacy, there must be an actual legislation. Second—that legislation must stipulate a State need and there is enough jurisprudence to disqualify needs that are frivolous or not entirely necessary, based on the objects towards which that department of government is working. Third—the collection or processing of data must, in relation to the stated need, be proportionate. Once again there is no way to legislate as to what is proportionate processing, but there are enough judicial precedents to guide the court on a case by case basis.'

During the course of the case, the government also announced a committee under Justice B.N. Srikrishna to come up with a data protection law. Over the next months, the committee would organize a series of public consultations to draft the law. The questions now went far beyond Aadhaar, and looked at data protection at a fundamental level, including how journalists should treat information and data.

Aadhaar, in effect, had triggered bigger debates around privacy, state capacity, cybersecurity, and exclusion. As the scope widened, it drew more people into the arena.

Geek Heresy

In many ways the divide between the development activists focused on exclusion and the technology activists focused on privacy, security, and fraud was an unnatural one, and should not have existed in the first place. Before he founded HasGeek, Jonnalagadda had spent a few years working at the intersection of technology, government, and society. At Comat Technologies, a tech-focused social enterprise, he got hands-on experience developing a solution for Karnataka's PDS. His big takeaway from that project was this: Technology can end up influencing, even dictating the way a real-world system works. (In that way, technology is politics). But at the same time, technology does not solve any problem by itself. Without intent and capacity, technology can make things worse. The problems of exclusion in places such as Jharkhand and Rajasthan post implementation of Aadhaar-enabled solutions were results of that. The impact of such solutions are almost always felt by the poor.

The lack of intent and capacity in the system as a whole led to issues that were the concerns of the elite—privacy, security, and surveillance—and they were the very same factors that led to exclusion. The two streams had to converge sometime.

And they did ahead of the biggest battle of Aadhaar in the courtroom.

In all these, the way UIDAI had been responding to these criticisms left a lot to be desired. When activist Sameer Kochhar posted a video to show how the system was prone to replay attacks—using stored biometrics—UIDAI responded by filing an FIR against him. Similarly, *Tribune* newspaper reporter Rachna Khaira faced the same fate when she revealed that credentials that gave access to demographic data of Aadhaar users could be bought for as little as Rs 500.[9] It was seen as assault on freedom of speech, and as a reflection of UIDAI's inability to plug those leaks.

Through all these, the Congress, the political party that rolled out Aadhaar, sent mixed signals. In a tweet in January, Rahul Gandhi, the president of the party, distanced himself from the project, saying, 'UPA's Aadhaar = A voluntary instrument to empower citizens. NDA's Aadhaar = A compulsory weapon to disempower citizens.' A few weeks later, his party, which is in power in Karnataka, passed a bill making Aadhaar compulsory for availing subsidies and benefits from the government. Similarly, Kerala, ruled by the Communist Party of India (Marxist), has been rolling out programmes linked to Aadhaar, even as the party general secretary Sitaram Yechury has been criticizing the programme.

During a panel discussion at the Bangalore Literature Festival in 2017, one of the authors of this book, Charles Assisi, quizzed Congress leader Jairam Ramesh on why politicians—including himself—keep changing their position on Aadhaar. Ramesh, for example, was one of the first ministers to launch Aadhaar-enabled applications in the UPA regime, and played a major role in coming up with the slogan, *Aadhaar—Aam Aadmi ka Adhikaar* (Aadhaar—the right of the common man). Yet, he also played an important role in stopping the rollout of DBT ahead of elections. Ramesh elaborated on some of the concerns regarding the scope of Aadhaar. But there was one sentence that received enthusiastic applause: 'Where I stand depends on where I sit.'

Towards the end of the panel discussion, Ramesh decided to take an instant poll on Aadhaar. How many of you think Aadhaar is a good idea, he asked. The audience was mostly liberal, and one of the organizers had indicated that many would be hostile to Aadhaar. But, in response to the question, over 60% raised their hands in favour of the idea. Ramesh was unfazed. He followed it up with another one. How many of them thought the project was implemented badly, he asked. 80% raised their hands. With that the discussion ended.

Kesavananda Bharati 2.0

When Ramesh left the stage at the Bangalore Literature Festival, he looked pleased at the way the discussion ended. But he was also setting up for a bigger, far more crucial fight. This one was to be played out, not in the pleasant climes of Bengaluru, but in the blistering heat of Delhi at the Supreme Court. There, he was among the petitioners questioning the constitutional validity of Aadhaar.

The case against Aadhaar was led by his fellow petitioner, retired Karnataka High Court judge Justice K.S. Puttaswamy, whose team of lawyers had successfully argued in 2017 that privacy was a fundamental right. Now, his case was that Aadhaar violated that fundamental right. Ramesh was adding weight to that argument by saying that Aadhaar should not have been passed as a money bill (which didn't need the approval of the upper house of the Indian parliament). P. Chidambaram, who was himself under political heat, argued his case.

The hearing started on 17 January 2018, and went on for 38 days, till May 10. Only one case in the history of Indian judicial history went on longer than that, and that was *Kesavananda Bharati v. State of Kerala* back in 1973. It went on for 68 days, led by eminent jurist Nani Palkhivala, and established that parliament cannot pass any law that violates the fundamental rights. Widely cited as one of the landmark cases that changed India, it made Bharati, the pontiff of the Edneer Mutt, nationally famous.

Shyam Divan, Kapil Sibal, Gopal Subramanium, K.V. Vishwanathan, P. Chidambaram, and Arvind Datar might well have felt the same sense of history as they made their case for why the Supreme Court should strike down parts of or the whole of the Aadhaar Act.

Their arguments were covered in the media day after day. And if anyone had any doubt about the risks of Aadhaar, the papers and social media would have cleared them. The lawyers' arguments in court and the news breaks on Aadhaar coincided—underlining the fact that they were not debating arcane points of law, but issues that people faced on the ground day after day.

There were primarily three big messages.

The first, that the Aadhaar ecosystem is leaky.

In January, a couple of weeks ahead of the Supreme Court hearings, a Jalandhar-based reporter of the *Tribune* published a story that said for Rs 500, one could get access to details of any of the more than 1 billion Aadhaar numbers. By most accounts, it took UIDAI by surprise. The body reacted by issuing denials and insisted (correctly) that the core biometric database was safe. It then filed a case against the reporter. It had done the same in two other cases, with limited backlash.

This time, however, it grabbed international attention. The news about the leaky nature of the ecosystem got a huge boost when a 'French' hacker, popularly known as Elliot Alderson, and a Twitter user with the handle @fs0131y simultaneously started pointing out how vulnerable the Aadhaar ecosystem was. They pointed to websites and apps with security holes only amateurs could leave. These leaks, they pointed out, violate privacy and could lead to fraud.

The second argument is that Aadhaar leads to widespread exclusion. During the hearing of the case, petitioners urged that the court should direct the government to compensate those who were excluded from welfare schemes because of Aadhaar.

A tragic case that was highlighted was that of Santoshi Kumari of Jharkhand. The 11-year-old died of hunger allegedly because she was refused ration on account of not linking her Aadhaar number. It was held up as an example of what exclusion can do to the poor.

A survey by IDInsights, published just after the hearing ended, showed that 0.8%, 2.2%, and 0.8% of PDS beneficiaries in rural Andhra Pradesh, Rajasthan, and West Bengal, respectively, are excluded from their entitlements due to Aadhaar-related factors. In percentages, these look minimal. In absolute numbers though, in a country as large as India, it means about 2 million individuals every month. (Exclusion due to non-Aadhaar-related causes such as non-availability of rations, was much higher. In Rajasthan and West Bengal, it was 6.5% and 5.2%, respectively. In Andhra Pradesh, it was 0.3%).[10]

The third argument was the risk of poor implementation or plain misuse. The argument in court questioned some of the widely held assumptions about Aadhaar, one of which was that the Supreme Court explicitly wanted all customers to link their cell numbers to Aadhaar.

It turned out that the court had passed no such order. It was interpreted to be so by TRAI, which had in turn asked all telecom companies to make Aadhaar mandatory.

Even as the hearings were going on, all news on potential or real misuse of data by countries such as China (spying on its own citizens) or Russia (which allegedly used data to rig the outcomes of American elections) were used to look at what could go wrong with a digital infrastructure like Aadhaar.

The lawyers representing the government made their best case for why the project is a viable one. Their arguments though were dismissed by vocal social media activists and went mostly unreported in the mainstream media. The asymmetric reporting might have had to do with the fact that their arguments were broadcast from handles of lawyers assisting the petitioners' counsel.

On its part though, the government went into overdrive and began a scenario-planning exercise. People we spoke to say that the PMO wanted to be briefed regularly on where things are headed and what the prognosis looks like.

While the team at work on the plan looked at multiple options, two were at the top of the list. The first, and the worst case scenario, as they saw it, was that the court could strike down the Aadhaar Act itself, either because it violates an individual's fundamental right to privacy, or because it was passed as a money bill. If this happened, the plan was to come up with a new law.

The second, and what everybody reckoned was a more likely scenario, would be that the Supreme Court would allow the government to use Aadhaar for its subsidy programmes. But it would place restrictions on how may it be deployed for private use.

UIDAI believes there are technical solutions, such as virtual IDs, that will answer the potential concerns.

As the book goes to press, the Supreme Court is yet to give its final judgment.

The long hearing in the court—and the arguments outside it—pitted one group against the other, and often it looked as if they were scorching the earth in between instead of finding a common ground. Yet, for cautious optimists, these fights are much needed to find a balance between the different needs of the different groups. The cacophony will eventually evolve into some kind of harmony that benefits all.

A Common Ground

Economist Amartya Sen has argued that famines don't happen in democracies because of the feedback mechanisms in place. The state can get to know of the negative news on the ground fast and take action. 'If profit is the oxygen for a business, criticism is the oxygen for the government,' Ashok Pal Singh of India Post Payments Bank told us.

Dissent and criticism of public policies have had a long tradition in India, and the country is a better place for that. Criticism on Aadhaar follows that tradition. It's easy to see why critics, both from the social sector and technology, have latched on to it. Aadhaar has literally touched a billion lives. It is impacting the way the government, business, and society work.

It has also gained international attention—from Bill Gates to Paul Romer, from the World Bank, and a number of countries, both developing and developed.

At the Bengaluru event, Jairam Ramesh said that India is the land of the Buddha, and taking a middle path on Aadhaar would be a wise thing to do. The audience seemed to like it. But one only has to listen to the loudest voices on Aadhaar to know that that has hardly been the case. In fact, the title of the panel discussion at Bangalore LitFest captured the polarization well. It asked if Aadhaar was utopia or dystopia.

While many of our conversations with many of the critics were nuanced and displayed an understanding that implementation of public policy is far more difficult than running campaigns, somehow on public platforms they seemed to be bereft of those qualities. At the same time, our conversations with supporters of Aadhaar revealed that they understood and deeply cared about the issues that the critics raised—those of exclusion, misuse, and faulty implementation—and yet in public debates they tended to downplay those concerns.

At a panel discussion on Aadhaar and India Stack at Bengaluru Tech Summit, there was a question on the risks of these platforms. The representatives speaking for India Stack didn't raise a single issue that critics were crying hoarse about.

There was also a third set we encountered, people who were involved in the design and development of Aadhaar and India Stack. They cared as deeply, if not more about these issues. They were also the doers, and were struggling hard with various government agencies to improve the system, in terms of the design of applications built on Aadhaar and their implementations. But it has not been an easy task. In 2017, UIDAI announced that it would offer a Virtual Aadhaar option to users, as a way to reduce the risks to privacy. It was reported as an impact of activism and the ongoing court case. In fact, Pramod Varma has been pushing for it right from 2010. One person who has worked with the governments in the past said, 'It's not easy for individuals to convince the system that things have to change. It can take years. Some activists seem to believe that sending a few emails would do the trick. It never does.'

That critics don't understand the complexity of the system and the difficulty of navigating it is not the only complaint supporters have about them. They can be broadly divided into three categories.

One is that they focus only on Aadhaar, when the underlying issues are much more widespread and possibly far more serious. For example, exclusion, poverty, and deprivation are big issues across the country. Right in Delhi, the country's capital, nearly 50 people died of cold in a matter of a week in December 2017. Yet, many critics give the impression that there were few problems on the ground before the implementation of Aadhaar, refusing to acknowledge the positive impact it has had. They point to its weaknesses in the present phase, calling for its wholesale destruction, almost as if doing away with the solution will take care of the problem it was meant to solve. There's no denying that Aadhaar, when implemented well, can make lives better for a large majority of the downtrodden.

Similarly, they talk about privacy and surveillance as if Aadhaar poses the greatest threat, ignoring the fact that the government runs at least five programmes that are explicitly designed for surveillance.

In fact, a vocal critic who considers Aadhaar a surveillance programme argued that Crime and Criminal Tracking Networks and Systems (a government-run surveillance programme)[11] is an 'egovernance infrastructure for law enforcement agencies', and insisted that no one can say anything about NATGRID (yet another surveillance programme)[12] 'with certainty'. As most of us are in the dark about these programmes, we cannot call the government out about data interception and overreach. (Aadhaar poses these risks only when it is misused or its data is leaked.) The result is the impression that Aadhaar is the only project we need to worry about.

Many dismiss the risks to privacy from smartphones, search engines, and social media, arguing that it's voluntary, ignoring the fact that consent is often broken, and the voluntary nature of a service does not guarantee that data will be safeguarded. To evaluate the risks, it's not enough to just look at voluntariness, it's also important to look at the quantum of data we give to others and the forms of data. Whether we like it or not, we are living in a digital world. And many of these issues can be resolved if our data protection laws are sorted.

However, to the extent that criticism of Aadhaar expands awareness and encourages action to cover the bigger issue, the critics are playing an all important role. For example, data protection laws—which were triggered by Aadhaar—will not be limited to just Aadhaar, but can potentially establish a data protection regime that others can set as a precedent for the rest of the world.

Two, they are too one-sided, negative, and sensationalist, spreading fear, uncertainty, and doubt when they should focus on bringing clarity. By doing so, they often undermine the cause they are fighting for, because it gets hard to separate signal from noise, and even genuine criticism gets dismissed. It's the job of the critics, activists, and advocates to make a strong case for their position. However, we are living at a time when the distinction between activists and journalists is blurring (as columnist Manu Joseph points out). When that distinction blurs, the line between news and propaganda blurs too, whether it is published in a newspaper or on social media, or gets discussed in television debate. For a typical media consumer—which is every one of us—it's hard to differentiate news from propaganda because many of the heuristics that we have traditionally used to identify propaganda (if it's for government or business it tends to be propaganda) no longer works in a complex world.

The Aadhaar critics' ecosystem lacks a voice that calls out errors and misrepresentations.

'I've been following and writing about the Aadhaar programme for the last few years and in my experience the debate is extremely polarized. Criticize aspects of Aadhaar and a legion of Aadhaar critics will praise you, but point out a flaw in their arguments and you get pilloried with ad hominem attacks,' Prashant Reddy Thikkavarappu wrote in media site *Hoot*.[13]

Thikkavarappu went on: 'I suspect the troll army of Aadhaar critics is also one of the reasons that there are so few within the community of Aadhaar sceptics who are willing to call out some of the inaccurate assertions being put out by their fellow travellers.'

The result of all this has left a section of population suspecting even valid criticisms against Aadhaar. It can be disastrous for the democracy.

Three, there is a missing segment among the critics—critics who also have the capability, willingness, and patience to change the system at local levels.

A critic's job is only to criticize, and it is unfair for a society to demand that they also solve the problems that they are complaining about—just as it is unfair for us to expect a film critic to produce a film that doesn't have the flaws he or she finds in others' movies.

However, for the system as a whole, there is a need for a set of sceptics who also have it in them to find solutions. A government might get too complacent. The critics might get too impractical, because they mostly deal in the world of theory. Men of action can get frustrated with critics. As Theodore Roosevelt said:[14]

> It is not the critic who counts; not the man who points out how the strong man stumbles, or where the doer of deeds could have done them better. The credit belongs to the man who is actually in the arena, whose face is marred by dust and sweat and blood; who strives valiantly; who errs, who comes short again and again, because there is no effort without error and shortcoming; but who does actually strive to do the deeds; who knows great enthusiasms, the great devotions; who spends himself in a worthy cause; who at the best knows in the end the triumph of high achievement, and who at the worst, if he fails, at least fails while daring greatly, so that his place shall never be with those cold and timid souls who neither know victory nor defeat.

However, there is still space for a set of pragmatists—who are as frustrated about the weaknesses of the system as the staunchest critics are. But instead of calling for the destruction of the infrastructure or suspending it till it rots, they roll up their sleeves and improve the system.

As we did our reporting, we became less and less convinced that the two warring factions are those who are for and against Aadhaar. Instead we saw that the fight was assuming religious shades. On one side we saw those who believed that it is possible to create a perfect system, irrespective of all the flaws they might find in this world, in their own lives and in themselves. They wanted Aadhaar to be perfect, flawless. And with every imperfection they see in the system, their conviction that it should be destroyed intensified. It must be said that a good part of the blame must also go to the signals that they got from the government, that it has little interest in addressing the criticisms. On the other side, we saw the evolutionists. They were ready to accept flaws and imperfections in the system, as long as there is some movement towards removing them.

That's a reason why arguments about Aadhaar are likely to continue long after the fight in the Supreme Court has ended. We saw that Aadhaar itself is just a Lego block in a technological solution; and technology cannot solve a problem without systemic changes in law, economy, and society—or without the intent and capacity of people in power.

Notes

1. Utkarsh Anand, 'Would Like to Be Known as Nationalist Judge Rather than an Aadhaar Judge, Says Justice DY Chandrachud', *News18*, 1 February 2018, https://www.news18.com/news/india/nationalist-judge-rather-than-an-aadhaar-judge-sc-judge-dy-chandrachud-1648261.html, viewed on 11 July 2018.
2. 'Judge Raps Aadhaar Petitioners', *The Hindu*, 1 February 2018, http://www.thehindu.com/news/national/judge-raps-aadhaar-petitioners/article22625042.ece, viewed on 10 July 2018.
3. Anupam Saraph, 'How Does Aadhaar Threaten You?' *ET Tech*, 11 April 2017, https://tech.economictimes.indiatimes.com/catalysts/how-does-aadhaar-threaten-you/2277, viewed on 10 July 2018.

4. Jean Drèze, 'On Research and Action', *Economic and Political Weekly*, 2 March 2002, nirman.mkcl.org/images/downloads/articles/G_On_Research_and_Action.pdf, viewed on 10 July 2018.
5. Jean Drèze, 'Unique Facility, or Recipe for Trouble?', *The Hindu*, 25 November 2010, http://www.thehindu.com/todays-paper/tp-opinion/Unique-facility-or-recipe-for-trouble/article15715442.ece, viewed on 10 July 2018.
6. Mark Zuckerberg, 'Is Connectivity a Human Right?', Facebook, 20 August 2013, https://www.facebook.com/isconnectivityahumanright, viewed on 10 July 2018.
7. PTI, 'Right to Privacy Not a Fundamental Right: Centre Tells Supreme Court', NDTV, https://www.ndtv.com/india-news/right-to-privacy-not-a-fundamental-right-centre-tells-supreme-court-784294, 23 July 2018.
8. Amber Sinha and Srinivas Kodali, '(Updated) Information Security Practices of Aadhaar (or Lack Thereof): A Documentation of Public Availability of Aadhaar Numbers with Sensitive Personal Financial Information', The Centre for Internet and Society, 1 May 2017, https://cis-india.org/internet-governance/information-security-practices-of-aadhaar-or-lack-thereof-a-documentation-of-public-availability-of-aadhaar-numbers-with-sensitive-personal-financial-information-1, viewed on 10 July 2018.
9. Rachna Khaira, 'Rs 500, 10 Minutes, and You Have Access to Billion Aadhaar Details', *Tribune*, 4 January 2018, http://www.tribuneindia.com/news/nation/rs-500-10-minutes-and-you-have-access-to-billion-aadhaar-details/523361.html, viewed on 10 July 2018.
10. State of Aadhaar Report 2017–18, IDInsights, http://stateofaadhaar.in/about-state-of-aadhaar/, viewed on 10 July 2018.
11. Maria Xynou, 'Big Democracy, Big Surveillance: India's Surveillance State', Centre for Internet and Society, 28 February 2014,

https://cis-india.org/internet-governance/blog/big-democracy-big-surveillance-indias-surveillance-state, viewed on 28 July 2018.

12. Software Freedom Law Centre, 'No Escaping the Surveillance State?', *The Hoot*, 4 September 2014, http://www.thehoot.org/free-speech/privacy/no-escaping-the-surveillance-state-8742, viewed on 27 July 2018.
13. Prashant Reddy Thikkavarapu, 'Scaremongering over HIV and Aadhaar', *The Hoot*, 19 November 2017, http://www.thehoot.org/media-watch/media-practice/scaremongering-over-hiv-and-aadhaar-10395, viewed on 10 July 2018.
14. Theodore Roosevelt, 'Citizenship In A Republic' speech, Sorbonne University, Paris, 23 April 1910.

6.1

Who's Afraid of Aadhaar?

A Directory of the Pain Points of the World's Largest Identity Programme

DURING OUR REPORTING ON AADHAAR, WE SPENT a lot of time discussing its shortcomings and those of India Stack with the group that built it, its champions, and its critics. There are many groups within each of these broad groups. For instance, technologists and bureaucrats among the builders, businessmen and policymakers among its champions, tech and development activists among its critics, whom we discussed in this chapter. By no means are these the only ones. Manu Joseph, in his column in *Mint*, said that Nilekani himself divided the critics into four groups—'the privacy' gang; 'the-rights-of-the-poor' gang; 'the oh-my-god-1984-has-arrived' gang; and 'the Luddites', who are scared of technology.[1]

While that's a useful framework to understand the broad issues, we also wanted to understand the critics' concerns, complaints, and demands. Not surprisingly, these did not come from the naysayers alone, but also from its supporters and those who designed and built the system. A bureaucrat who was involved with Aadhaar, for example, was worried about making it mandatory. The team built the system on the assumption that it would be voluntary—and Nilekani, as recently as July 2018, in an interview with Vir Sanghvi, said that he doesn't believe it should be mandatory. The technologists who built the system were naturally concerned about security and bad implementation of solutions that linked to Aadhaar. There were also concerns that had more to do with emotions than reasons. But when we are dealing with humans, emotions matter too.

In the end, we tallied more than fifty reasons why people criticize Aadhaar. We've compiled it here. It starts with issues around its impact, for that's what most are deeply concerned about, and goes on to address its intent, its need, its solution, implementation, processes, and ecosystem.

Impact

Aadhaar has led to fraud, exclusion, security and privacy concerns, overdependence on an unreliable system, and exaggerated claims of savings.

- That the Aadhaar enrolment system can be manipulated to give fake IDs became evident even in the early days, when a man got his dog, Tommy Singh, enrolled for Aadhaar. Someone else got an Aadhaar for Lord Hanuman. Frauds got more sophisticated over time. In Uttar Pradesh, for example, police caught a gang that was using a hacked client application to enrol users.
- The strength of Aadhaar authentication is supposed to be its biometrics. A researcher showed that the system is prone to replay attacks (your biometric can be stored in a

device and be used to authenticate). UIDAI came up with a solution that makes such attacks difficult, but it also filed a case against Sameer Kochar, the researcher. The system is still prone to other kinds of manipulation, including fake fingerprints. With the prevalence of high-definition cameras, and the progress of 3D printing, it shouldn't be long before replicating fingerprints is child's play.

- One of the earliest fears that social activists had was that linking Aadhaar to subsidy schemes meant that the system would treat all those who do not have the identity as 'ghosts'—fake accounts leeching the system—and simply remove them as beneficiaries. One would have thought that with more than 1.2 billion people enrolled for Aadhaar this problem would go away. However, surveys show that to this day many of those considered ghosts are real people. Genuine beneficiaries also get excluded, because the linkage with the schemes fail for various reasons.
- In some cases, the Aadhaar-based solutions are designed so badly, they exclude people at the last mile. In one instance, in a departure from the earlier system, where anyone with the document could buy the rations, the system allowed only the head of the family to take rations from PDS shops. The head of the family might be unwell, out travelling, or at work. Such badly designed solutions can end up excluding people. In other cases, there were no backups to biometric authentication, were it to fail.
- Beneficiaries are excluded as they are unable to authenticate because of faulty devices, bad networks, etc. As Reetika Khera puts it: '… each month, people are being forced to cross five

meaningless hurdles in the form of electricity, functional PoS, connectivity, servers, and fingerprint authentication in order to access to their ration. Failing any one hurdle even once causes anxiety in subsequent months. Think of the ATM running out of cash, post-demonetization, just when it was your turn.'[2]

- Aadhaar doesn't solve all the problems. Often, it becomes another excuse for the system not to give the entitlement to the beneficiary.
- That the core database of Aadhaar has not been breached so far is no guarantee that it will not happen tomorrow. Right now, UIDAI expects people to take its word that it is safe and secure. However, going by its responses to previous breaches in the ecosystem, and by the lack of visible best practices, it's doubtful that UIDAI has the capacity to keep its database safe. Especially at a time when the battleground is shifting from the material world to the cyberworld. Countries such as the US have been investing significant resources on cybersecurity. There is no evidence that India is doing enough to protect its data.
- The government continues to claim—without evidence—that Aadhaar has helped it save billions of dollars. In fact, publicly available data suggests that the cost savings from Aadhaar have been minimal.
- Aadhaar is likely to lead to intrusion by the government (and even businesses) into our private lives, violating our fundamental right to privacy.
- All the nudging and enforcement to use Aadhaar is likely to increase government dependence on technology—before the infrastructure is capable of taking that load.

(Most of these concerns are shared by both critics and supporters. While supporters believe these can—and will—be fixed, critics argue that in some cases they cannot be resolved, and in others, that they will not be.)

Intent

- Some critics suspect the intentions of the ideators, designers, and executors of Aadhaar—whether government or private. That the government basically wants to empower itself, have more control over the people, and that Aadhaar is primarily a mass surveillance system. Moreover, even if the intentions are good today, they might turn bad tomorrow. As Sunil Abraham, a long-time critic of Aadhaar puts it, 'the original noble intentions of the Aadhaar project initiators will not constrain those in the future that want to take full advantage of technological possibilities'.[3]
- Similarly, others believe that the private players, starting from Nandan Nilekani onwards, were in it for money, power, or at least to get access to people in power. This view was seen in an *Indian Express* article[4] that asserted, 'executives who have worked or are working with the Unique Identification Authority of India (UIDAI)—the parent agency for Aadhaar—are launching companies or funding start-ups that offer Aadhaar-based services and products for a fee.'

(To this set of criticism, Aadhaar supporters say that questioning intentions often leads to a confirmation bias. That is, in a complex world, any decision or action can be considered bad if one starts with the assumption that intention is bad. It also leads to the other party questioning the critics' intentions.)

- The problems that Aadhaar promises to solve don't exist in the first place. Aadhaar supporters assert that Aadhaar is a detergent that can clean up databases. But is there really such a problem? Take PAN cards, for example. Government says that by linking PAN to Aadhaar, fake cards can be weeded out. However, according to an article in the *Wire*, 'out of 25 crore [250 million] PAN cards, only 0.46% or 11.57 lakh [1.157 million] are actually duplicate. These have already been cleared from the system. Furthermore, the total number of fake PAN cards the government has found is 645.'[5]
- The rationale behind Aadhaar is that lack of identity is the root cause of many problems with public goods and targeted subsidies. While problems exist, often they have nothing to do with identity. For example, analysing the midday meal scheme in the country, a Huffington Post[6] piece found that '[w]hat the scheme does not suffer from is beneficiary fraud, of the sort that even the Public Distribution Scheme (PDS) suffers to an extent'. Or as Reetika Khera[7] pointed out, 'the key question with respect to identity fraud (and the Aadhaar project) is what Senior Advocate Arvind Datar asked the government in court (during the PAN-Aadhaar linkage case), "Did you do any study?" The fact is there is no reliable evidence on the scale of identity fraud in welfare programmes.'
- Aadhaar was sold to the government and then to us by asserting that many Indians do not even have an identity. One way to find out if this is true is to see how many got their Aadhaar number by producing other existing

identification documents versus those who did not have any identity and had to rely on the 'introducer system'. It turns out that the government said that till 2016, when over 105.1 crore [1,051 million] residents had enrolled, only 847,366—or 0.08%—got Aadhaar through 'introducer system'.

- Some countries are doing well without a national ID. Why not India? We have to fundamentally question the need for a national identification like Aadhaar, when we know that advanced countries such as Canada and Australia are doing well without national ID programmes.

(To this set of criticisms, Aadhaar supporters typically point out that the need for an identity is increasingly being recognized as important by countries and multilateral development agencies, and that critics use bad data and/or bad reasoning to assert their claims.)

Solution

- Aadhaar is the product of techno-utopianists, solutionists. But, that's not a good approach, like Evgeny Morozov argued in *To Save Everything, Click Here*.[8] Many of India's problems can't be solved by technology. They need social, political, and economic solutions. Bringing in technology to solve such a problem is like bringing an axe when we need a screwdriver. Aadhaar, however, assumes that technology is a solution.
- A centralized database is more prone to getting hacked; there is a single point of failure; and it incentivizes bad actors to attempt to attack. The impact will be huge. There is a second-order impact too: It centralizes power—UIDAI comes under central government after all—and to that

extent tilts the balance of power between the Centre and the states.

- Supporters of Aadhaar often use the word 'open', hoping that some of the values that the open source movement stands for will stick to it. However, Aadhaar violates those principles, and was developed in a closed way. It has made it hard for others to scrutinize its code.
- Aadhaar was built as a platform, upon which other solutions can be built. However, UIDAI takes no responsibility for the mess that happens on the ground. It's power without responsibility.
- Biometrics is the wrong choice. Aadhaar should not have gone for biometrics because, unlike a password, you can't change it; once lost, it's gone. [If your biometric is compromised—if someone 3D-prints your finger—you are compromised permanently, because you can't reset biometrics].
- Biometrics are also probabilistic, which means there will be false positives and false negatives. This makes it perpetually vulnerable to both exclusion and fraud. A password, on the other hand, is either 'yes' or 'no'.
- Large-scale biometrics might not work. We don't know how good it is. UIDAI does not give the data. We should learn from other countries and ask why the UK abandoned its biometric ID project. In fact, democratic countries don't have biometric IDs, according to an analysis by two researchers.[9] They asked if countries with a poor democratic record are more likely to mandate an Aadhaar-like ID—and found out that was indeed the case.
- Making Aadhaar online only is a bad idea, because it leaves too many chances to external factors, such as Internet

connection, third party devices, websites, and so on. Smartcards would have been a better alternative—they are a more established technology. They can also sidestep some of the problems related to the centralized storage of biometrics.

- Aadhaar doesn't establish citizenship. Just about everyone—from Pakistanis and Bangladeshis to expatriates are getting Aadhaar. And therefore it poses a security risk.
- When Aadhaar was being rolled out, it was promoted as a voluntary programme. However, by linking it to PAN and other essential services, it has now become mandatory.

(To this, Aadhaar supporters point out that there are positives and negatives for all design decisions, and that critics tend to point out only the negative aspects, ignoring the positive.)

Implementation

- Enrolment should not have been implemented this way. UIDAI wanted to scale up fast. However, it has sacrificed quality at the altar of speed. It used private players for enrolment, which is wrong in the first place. The selection of enrolment agencies was not rigorous. As a result, too many rotten apples entered the system, and they had to be suspended. The applications developed for enrolment were not secure enough, and it emerged that they could be hacked.
- The defective enrolment process has meant that there's a large unverified database. And an unverified database is essentially useless.

- Applications designed by various state and district administrations are faulty. For example, the Telangana government designed its PDS in such a way that only the head of the family can buy from ration shops.
- The government has been pushing Aadhaar even in places where there is no compelling reason to do so. For example, for death certificates, for mobile phones, for driver's licence, for bank accounts. It was supposed to be mandatory only to avail government services.

(Many of these concerns are shared both by the supporters and critics. Supporters believe they will be solved. Some of the overuse of Aadhaar can be solved by federated IDs. For example, if PAN cards are deduplicated using Aadhaar, they can be considered as unique as Aadhaar. Similarly, the roll-out of virtual Aadhaar numbers should also address some of the criticsms. Critics are more pessimistic and say it might be too late.)

Processes

- India did not have an Aadhaar Act for a long time. A programme that is as disruptive as Aadhaar came into being without debate in parliament. In fact, the Aadhaar Act was passed well after enrolments, during the NDA regime.
- Even in the NDA regime, Aadhaar was passed as a money bill, which meant it did not have to be passed in the Rajya Sabha.
- The government has constantly flouted Supreme Court orders, by pushing Aadhaar in various schemes, even

though the Supreme Court had passed orders limiting the mandatory use of Aadhaar.

- UIDAI didn't do enough public consultations (especially technical). It's not open to outside expertise. It doesn't organize bug bounties to ensure that its security measures are good.

Ecosystem

- Aadhaar has been thrust on a country that is not ready—technologically, legally, or knowledge-wise, especially with respect to data protection, security, and privacy.
- Many government agencies have a cavalier attitude towards handling user data. Even where they want to treat it carefully, they do not have the capacity to do it. Sometimes, just a Google search is enough to access the data. In some cases, basic hacks are enough to access the entire database (as happened in the case of the Telangana state government). As the article in the *Tribune* showed, data can be bought for as low as Rs 500. In effect, while the biometric data might be secure in the central data depositary, the ecosystem is so leaky that sharing Aadhaar numbers anywhere is not safe.
- Besides problems with Aadhaar law—only UIDAI can raise complaints—we don't have the infrastructure to provide justice to potential victims.
- A majority of people are new to the digital world, are not aware of the risks, and might get conned. Aadhaar is being thrust on people without giving them enough time to learn the risks and the responsible use of their own data. It's not just about the users—who give away their Aadhaar numbers

and other sensitive information—it's also about the data handlers, who often insist on photocopies of Aadhaar. This often defeats the very purpose of online authentication.

(This is another area where Aadhaar supporters and critics are on the same page. The differences are in how to respond to these issues. Supporters insist that these issues must be fixed as we go along. Critics insist that the programme needs to be either destroyed or suspended till these are fixed.)

Given how often and how deeply critics and supporters agree on some of the most important issues related to Aadhaar, one might wonder what the fuss is all about. Why should there be such a high degree of ill will?

One very obvious trend that we noted was that there has been very little dialogue between the two groups except in settings that exploit the differences. For instance, television studios, which thrive on high-pitched debates rather than reasoned discussions; in courtrooms, where each party tries to win at all costs; or in public squares, where ego ends up playing a major role.

Otherwise, there has been very little interaction among the warring groups. In effect, when supporters meet, they basically agree on everything and go on to finding ways to fix the system. The competition is really about who can fix the system faster. That explains some of the recent changes and upgrades in the Aadhaar ecosystem, including the launch of Virtual Aadhaar, UPI 2.0, banks taking up enrolments, and in fact, higher investments in cybersecurity across agencies. This has not always been positive, for some of them have indulged in exaggeration of Aadhaar's benefits, and playing down its risks, resulting in misinformation.

When the naysayers meet, they also basically agree on everything, and the competition is often in who can spread the message louder and more effectively. This has in fact brought a lot of awareness about the risks of not just Aadhaar, but also of a digital economy. Their role in pushing the government to take the risks seriously and work on legislations like data protection laws has been—and will remain—significant. At the same time, as in the case of Aadhaar supporters, their actions have also had unintended consequences, resulting in excessive paranoia, misinformation, and other risks pointed out in the chapter.

In short, the debates around Aadhaar are not driven just by intergroup rivalry—between the critics and supporters—they are driven, even to a greater extent, by intragroup rivalry, with supporters trying to outdo other supporters, and critics trying to outdo other critics.

Notes

1. Manu Joseph, 'When You Give Your Biometrics to Modi', *Mint*, 10 April 2017, https://www.livemint.com/Leisure/28JmBgGLgWLclLxQtzd9KI/When-you-give-your-biometrics-to-Modi.html, viewed on 23 July 2018.
2. Reetika Khera, 'Why ABBA Must Go: On Aadhaar', *The Hindu*, 13 November 2017, https://www.thehindu.com/opinion/lead/why-abba-must-go/article20353913.ece, viewed on 26 July 2018.
3. Sunil Abraham, R.S. Sharma, and Baijayant Panda, 'Is Aadhaar a Breach of Privacy?', *The Hindu*, 13 March 2017, https://www.thehindu.com/opinion/op-ed/is-aadhaar-a-breach-of-privacy/article17745615.ece, viewed on 26 July 2018.

4. Krishn Kaushik, 'Aadhaar Officials Part of Private Firms that Use Aadhaar Services for Profit', *Indian Express*, 5 October 2017, https://indianexpress.com/article/india/aadhaar-officials-part-of-private-firms-that-use-aadhaar-services-for-profit-4874824/, viewed on 10 July 2018.
5. James Wilson, 'Why Duplicate PAN Cards Are Not as Big an Issue as the Modi Government Claims', *The Wire*, 13 May 2017, https://thewire.in/economy/duplicate-pan-cards, viewed on 23 July 2018.
6. Rukmini S., 'There Is a Place for Aadhaar, but the Mid Day Meal Is Not It', *Huffpost*, 4 March 2017, http://www.huffingtonpost.in/2017/03/04/there-is-a-place-for-aadhaar-but-the-mid-day-meal-is-not-it_a_21873276/, viewed on 10 July 2018.
7. Reetika Khera, 'The Real Beneficiary', *Indian Express*, 2 June 2017, https://indianexpress.com/article/opinion/columns/uidai-aadhaar-card-the-real-beneficiary-4684994/, viewed on 9 July 2018.
8. Evgeny Morozov, *To Save Everything, Click Here* (New York: PublicAffairs, 2013).
9. Rohan Venkataramakrishnan, 'Are Countries with a Poor Democratic Record More Likely to Mandate an Aadhaar-Like ID?', *Scroll*, 29 September 2017, https://scroll.in/article/851282/are-countries-with-a-poor-democratic-record-more-likely-to-mandate-an-aadhaar-like-id, viewed on 10 July 2018.

7

India and the World

ON 1 FEBRUARY 2018, WHEN FINANCE MINISTER Arun Jaitley presented the Union Budget, one of his statements sounded rather innocuous.[1] But it was a very significant one.

> Technology will be the biggest driver in improving the quality of education. We propose to increase the digital intensity in education and move gradually from 'black board' to 'digital board'. Technology will also be used to upgrade the skills of teachers through the recently launched digital portal DIKSHA [Digital Infrastructure for Knowledge Sharing].

He then moved on to other things.

What he left unsaid, though, is that this comment was a precursor to larger things. Beginning the next academic year, upwards of 200 million textbooks in India will have QR codes printed on them—on average, one QR code per chapter, in five languages. These textbooks will be made available across government and private schools in Andhra Pradesh, Maharashtra, Uttar Pradesh, Tamil Nadu, and Rajasthan.

When a QR code is scanned on a phone, it brings up more information on the topic being discussed—including extra worksheets to practise further.

Printing of the textbooks has already commenced. To start with, 50 million children are being targeted. The plan is to reach 200 million by 2020.

For things to get this far, work had been initiated a long while ago at the Ministry of Human Resources and Development. This is a part of a project called the National Teacher Platform (NTP). It was thought up because those at the ministry could see an issue on hand. There are 10 million teachers across schools in India aided either by the central or state government. Data showed that only 15% of all teachers received training of any consequence after they joined the workforce.

A strategy and approach document released by the National Council for Teacher Education in May 2017 talks about how they plan to go about it:[2]

> A number of states across the country have expressed an interest in building technology platforms for teachers. Instead of each state building individually, the National Teacher Platform (NTP) will be built as a common public good.

This was, perhaps, the first time in India that a government body had spoken about building a digital public good. Much of the learnings around how to build and implement this had come from Project Aadhaar and India Stack.

The team that had worked on the identity project had disbanded. Some of them were at work on multiple projects at a not-for-profit called EkStep Foundation in Bengaluru. The learnings gleaned from implementing Project Aadhaar is something they have documented and are deploying over various domains—not just in India, but across the world.

How they are going about it is something that has got the attention of policymakers in India and other parts of the world. DIKSHA is one outcome. This, because they like the philosophical premise on which it is built. At EkStep, they call it 'Societal Platforms'.

'It is not something that you can objectify as a technology or a piece of software. It is a way of thinking,' explains Sanjay Purohit. He used to be with Infosys in North America before moving to EkStep.

A Societal Platform is built on the premise that the most precious commodity in the world today is data. So far, data is owned by the government or the business that collects it. In the framework that is Societal Platforms, though, the underlying premise is that data belongs to the individual. An individual may consent to share it with regulatory authorities to comply with norms or private entities if he or she thinks it appropriate. It is the same principle on which Aadhaar and India Stack are crafted.

When this was first articulated in policymaking circles, it caught the attention of a few global thought leaders. Paul Romer, chief economist at the World Bank until January 2018, was one among them. It compelled him to ask for a closer tour of India Stack and the outcomes that have emerged from it. Romer liked what he saw.

The 2017 edition of the Faster Payment Innovation Index,[3] an annual report compiled by FIS Global, a US-based think tank, placed India at Level 5, at the top of the 25 countries it surveyed. India was the only country at this level. This, because the think tank reckons India has the world's most evolved digital payment system and infrastructure that 'meets most features maximizing customer value'.

At Level 4 are the payment systems in Finland, the UK, Singapore, Denmark, and Switzerland. China stands at Level 3—which means, it has managed to build an infrastructure that 'meets most features enhancing customer value'.

The US and Canada do not make it to the Top 25 because in FIS Global's reckoning, while the infrastructure there is good, it still has to make more headway. Pretty much every country is grappling with many issues on hand.

South Korea, while a technologically advanced country, has had to build its systems ground up: from 2004 on, its national identity system started getting compromised. In 2014, it compelled the government to decide that the system must be revamped. Current estimates have it that with the plan it has, it will take at least a decade and billions of dollars to implement—all this for a nation that has just 50 million people.[4]

Denmark launched NemID in 2010. This is a credit card-sized identification system that is an improvization over what it already had. It took over three years to put in place. To get one though, a citizen must answer some check questions that are already logged into a central database; or bring a witness along who can attest to their identity. If a citizen declines to do either, the NemID code card will be mailed to their home address that was registered in a centralized database. What exists in South Korea and Denmark are thought of as rudimentary.

It sounds ironical that such systems exist in countries that are more advanced and place a premium on the value of data. Data can be mined in multiple ways to extract much information about an individual. When done on large numbers of people, one can influence public discourse.

The biometric identity system being built in Japan offers a pointer to what is possible. That the system is now antiquated, must be revamped to plug leakages, and offer greater benefits to citizens was reported as early as 2014 by the Financial Times.[5] The government is hoping that when completed, it will shore up tax revenues as well. Its implementation, though, has left citizens furious in a nation where privacy is considered sacrosanct.

The English language edition of Mainichi Newspapers[6] in Japan reported:

> The economy ministry is also planning to analyze the data provided by visitors using the system, including age group and gender, in combination with their purchase histories to determine consumption trends and the particular shopping tendencies of certain nationalities. The goal is to provide very specifically tailored hospitality to foreign visitors in the lead up to the Tokyo 2020 Olympic and Paralympic Games, and thereby encourage them to spend more in Japan.
>
> The ministry is aiming to conduct test runs of the biometric shopping system in each region of Japan over the next three years, with a private sector rollout in fiscal 2019 or later.

As far as private entities go, there is much to be gained by owning data. Mary Meeker's Internet Trends Report 2017 has it that in 2016, Google and Facebook captured 71% of all ad spending in the US—an 89% growth over the previous year. And mobile payments in China controlled largely by AliPay and WeChat hit $5.5 trillion, almost 50 times the size of America's $112 billion market.

These are the kind of numbers that tempt private entities driven by the profit motive to behave in perverse ways. It is something the world is waking up to. In February 2017, the nine largest banks in the UK were given a year by the regulatory authorities to develop and adopt an open-banking API for customer data. They have been compelled to comply.[7] Other regulators in the EU moved along similar lines. Confusion abounds,[8] and most have asked for an extension to comply with the new ask. They don't know how to go about it.

India has already begun to use an identity system of the kind many countries aspire to build, and it is operating at scale. That is why Project Aadhaar is getting a lot of attention from many governments.

Singapore is one such country. It currently has a national identity system called SingPass. While it has stood the government in good stead until now, Vivan Balakrishnan, the minister in charge of the Smart Nation Office, has acknowledged the system needs to be upgraded[9] after it was reported that data was getting compromised as early as in 2004. He is among those in the government there who think it is necessary that a new system be built and that it must include biometrics, encryption, and open APIs. What Singapore has, he went on the record to state, 'is not good enough' and a national identity system 'is absolutely essential if we are going to have secure transactions in the digital world.'

Sangeet Paul Choudary of INSEAD Business School and executive educator at Harvard Publishing School is part of the Singapore government's think tank. He has suggested to the authorities there that Project Aadhaar be looked at as a prototype along which a new system can be architected. He has been closely engaged with the founding team at UIDAI and likes the premise that drives Societal Platforms.

When asked why, his response was that it has much to do with how the world is changing. In the past, Singapore was strategically located to be a hub that facilitates trade. With all things going digital though, the leverage a location provides may not be a good enough strategic advantage. Much else must be done to retain the edge. One of the questions he has posed to the government: How do they work to position Singapore as a Platform Nation?

However, every country must think of Singapore as a 'digitally neutral' place where they feel comfortable conducting their trade. Societal Platforms is a construct that fits there. To get there, though, a new-age identity system must be put in place.

The 'Econo-mix' of Identity

There is no ambiguity across the world that a national identity system is much needed. Eve Maler offered some pointers to why that is so in a conversation with Bloomberg.[10] She is part of the leadership team at ForgeRock, a digital identity firm with Norwegian roots. 'What's happening is that when Aadhaar servers are saying "yes, this person is who they are saying they are", the question is, how much information comes along with that authentication event?'

That is indeed an interesting question. It leads to first principles and insists we ask: What is the significance of identity for any economy?

This is a question that can be debated at multiple levels and has been the subject of much study. George Akerlof, the Nobel-Prize-winning economist, had gotten into the many nuances of this as early as in 2010 in his book *Identity Economics*.[11] He described it as a 'fuzzy' idea that most economists have trouble wrapping their heads around because it lies at an intersection where economics meets sociology, anthropology, psychology, political science, and literature. But for a society to develop, all individuals who are a part of it must have a formal identity independent of the ones conferred by history, culture, etc. Akerlof says:

> ... in the real world, individuals' conceptions of fairness depend on the social context. In many places it is seen as fair and perhaps natural to treat other people in ways that elsewhere are considered unfair and even cruel. This observation is as important as it is obvious. In India, upper castes do not treat lower castes equally. In Rwanda, Tutsis and Hutus do not treat each other equally. In America, whites have not treated blacks equally.

After having gone over what he had articulated and listening to multiple voices such as these, the *New York Times* offered an opinion[12] at a time when the world was in a recession:

> ... an economy works well when people personally identify with it, so that their self-esteem is tied up with its activities ... Solutions for the economy must address not only the structural instability of our financial institutions, but also these problems in the hearts and minds of workers and investors—problems that may otherwise persist for many years.

As articulated earlier in this book, large parts of the world, India included, are economically underprivileged. But when India Stack is integrated into the economy, it lets more people participate, and a new economic engine, or model of development—like in education, or healthcare for that matter—can emerge.

'Data is now seen as a strategic resource as well by most governments,' explains Sanjay Anandaram, a venture capital investor and part of the core team at iSPIRT. 'Most governments are deeply suspicious of what may be the underlying motive of any platform that emerges from a country.' Anandaram suggests how countries now engage with each other geo-politically as well is determined by how platforms evolve.

What most people can see are large commercial platforms led by American entities Facebook, Amazon, Google, and Netflix, among others. They place a premium on gathering data. Authorities in China could see why a long while ago. That is why they decided to fund entities to capture data and created barriers to restrict American entities from China.

China has been at work as well to craft a meticulous strategy. One among the many outcomes has been the emergence of Baidu, Alibaba, and Tencent, also called the BAT trinity, to challenge the dominant American narrative.

The rest of the world must choose between these two narratives.

When Societal Platforms are presented as an alternative, governments, multilateral agencies, coalitions of philanthropists, and businesses see the emergence of a third narrative and are open to examining if there may be merit in it.

This is because Project Aadhaar shows how a platform is being used to ensure money pledged gets to the intended recipients, thereby plugging leakages. What they are also witnesses to is a malleable platform built on open source software and which can be crafted to suit the requirements of the constituencies they may want to serve.

Bill Gates was among the earliest philanthropists to connect the dots after listening to presentations by EkStep in Bengaluru in December 2017. EkStep was co-founded by Nandan and Rohini Nilekani with Shankar Maruwada.

That Gates was familiar with India Stack from his earlier visits to India helped. He and his wife Melinda spoke to the others at Co-Impact, an initiative that describes itself as 'a global collaborative for systems change'. Core partners at Co-Impact include the Rockefeller Foundation, and billionaires like Richard Chandler, Jeff Skoll, and Romesh and Kathy Wadhwani among some others.

Buy-in from people like these was very important, explains Sanjay Purohit, a former Infosys executive, who is now a strategy advisor to EkStep Foundation. This, he says, is because a collaborative initiative like Co-Impact brings together philanthropists who have the risk capital and are wedded to a purpose as well. If they see merit in an initiative that can catalyse innovation in the purpose they hold close, they have the capital to invest and allow people with the innovative idea to function like a start-up.

A government cannot take this kind of risk because taxpayer money is involved. No commercial investor would be willing to invest into a premise like this either, because the returns on capital are not visible. While the idea may interest an NGO, it may not have the capital to invest in it. That said, Purohit says, to make a Societal Platform come alive, all of these entities have to come in at some point. At different points in time, different skill sets are called for, he says.

To go back to Co-Impact, everybody who was part of the collaborative agreed there is merit in engaging with EkStep, and on 15 November 2017, an announcement went out globally that it would be a technical partner. This, Co-Impact conveyed, is because it is concerned about how the $500 million it has pledged to invest in education, healthcare, and in creating economic opportunities for the underprivileged is deployed.

To begin with, if the right kind of technology platform that could be used seamlessly across the world is in place, they could monitor it in real time, and commit more if need be.

The announcement did not go unnoticed. In the first week of December 2017, Karen Bradley, the British secretary of state for Digital, visited India with a team on a mission. The British government reckons their system must be fixed urgently. The population is ageing and the intended beneficiaries are not getting what is due to them.

They had identified the problem as early as 15 years ago and launched a programme called Verify to plug the loopholes. But to verify the identity of each individual then cost them £60. It was turning out to be a burden on the exchequer and unviable. If a system like India Stack can be exported to the UK, authenticating the intended recipients can be done in under $1. As things are, they need it done yesterday.

Kyrgyzstan articulated its interest in replicating the architecture and has asked what it'll take an Indian contingent to train its people to do it. Multiple inquiries have come in from the Middle East as well. It wasn't incidental that Mohammed bin Zayed Al Nahyan, the crown prince of Abu Dhabi and deputy supreme commander of the UAE Armed Forces, was invited to be the chief guest at India's Republic Day Celebrations in 2017.

It was Prime Minister Modi's way of sending out an overture to the UAE that the countries are friends and that there may be merit in exploring trade ties.

In the backroom, since the time Modi took over at the PMO, there has been much happening. As early as 2016, Russia, Morocco, Algeria, and Tunisia[13] were among the earliest countries that expressed interest in Project Aadhaar.

Early in 2018, Sri Lanka signalled its intent to replicate Aadhaar.[14] This was considered fairly significant from an Indian perspective. Sri Lanka had ceded control over the strategic Hambantota port[15] for 99 years in the first week of December 2017 to China after it failed to meet its obligations to Chinese companies working in the country. The island state was struggling with its finances. Indian diplomats were fretting because the port is quite close to Indian waters. How may it be possible to reach out to Sri Lanka? An answer emerged in the form of Project Aadhaar.

Called UMANG there, it aims to give Sri Lankan citizens a UID, access to various schemes, and help make education accessible.

More overtures have gone out from the PMO, and in a first of sorts, all heads of states from ASEAN countries were invited to the Republic Day Celebrations in January 2018. In all, at least 20 countries are in various stages of discussions with the Indian government to understand if there may be a way to create a native version of Project Aadhaar.

There is a reason these heads of state are being wooed.

Much like with Sri Lanka, there is concern in Indian diplomatic circles over China's growing dominance in the neighbourhood. Dialogue must be initiated, and offerings must be made. On the table right now is a demonstration of what's been accomplished on the back of India Stack—and an offer to collaborate with people on the ground in their countries. Once implemented, it will be handed over to the government to run as it deems fit.

An expectation is that Indian expertise will be used to maintain the systems, train the ecosystem there on how to use it better, and innovate in a local context. The Indian takeaway is that innovations cannot be imported and imposed on a society.

That is what China is attempting to do. This is a story many countries, particularly in Africa and South America, do not like, but is a bullet they feel compelled to bite. If India offers them a more benign narrative to craft a different economic trajectory, they are willing to align. Else, what happened in Sri Lanka can very well happen to them as well.

Tri-sector Athlete

Soon after their stints at UIDAI, the team disbanded. Nandan and Rohini Nilekani headed to the US to take a course on public policy at Harvard University. Their conversations there suggested India's demographic dividend can turn into a disaster[16] if things aren't fixed urgently. The most recent numbers and reports, for instance as compiled by IndiaSpend, offer a pointer to why.

- By 2030, India will have 1.03 billion people in working age.[17] That is just 12 years away.
- India added the fewest jobs in the organized sector across eight industries.
- 93% of Indians have no formal pay or social security. Of these 47% live in rural areas and are employed as agricultural labour.
- As many as 60% of those who live in rural areas and depend on agriculture for income do not find employment through the year.

There are multiple reasons for this. These include limited progress on building health infrastructure, providing basic education, and absence of the right set of job skills. What it means is that time is of the essence.

Multiple sets of leaders have been seeking answers to these problems since 2008. Among the more vocal of these is Dominic Barton, global managing partner at McKinsey.

Barton has spent considerable time talking to people across the world and asking what ought to be done. Some of his thoughts are captured in an essay he wrote on the nature of capitalism in the long term for the March 2011 issue of *Harvard Business Review*.[18] He laments the short-termism of Wall Street that led to its collapse in 2008 and argues that a new class of leaders is needed. That is why, he writes, 'Tomorrow's CEOs will have to be, in Joseph Nye's apt phrase, tri-sector athletes: able and experienced in business, government, and the social sector.'

What Barton articulated sounded familiar to those who had worked on implementing Aadhaar, including Shankar Maruwada, who had co-founded FourthLion Technologies, which helps politicians use data analytics. (It helped Nilekani during his campaign in 2014.) The first set of phone calls went out.

Sanjay Jain had gone back to the private sector as founding CTO of Novo Pay Solutions at Khosla Labs, an incubator created by Vinod Khosla, the billionaire co-founder of Sun Microsystems. It was being crafted on India Stack. This call, though, prompted him to join his old teammates. Pramod Varma's technical expertise was much needed at UIDAI and his association with the agency continues. Deepika Mowgli Shetty, who used to help Nilekani think through legal policy at Aadhaar, was among those who joined EkStep.

Purohit sounded like someone who could help with EkStep's stated mission to think up solutions to problems India must resolve urgently. The team had learnt much about the country and the developing world while they were at work on Aadhaar.

Purohit was part of the team at Infosys, was based in the US, and was among the key people instrumental in crafting a global strategy for the company. Even though he was not part of the team that worked on Aadhaar, he had seen it evolve from a distance.

He listened to the learnings from Project Aadhaar. Whatever is thought up must be scalable, sustainable, and able to create long-term impact. To that extent, they must attempt to build a platform.

In building Aadhaar, an entire ecosystem called India Stack had been created. Aadhaar, as we now know, is only that layer which authenticates an individual's identity even without their presence. It is only when all the other parts of the platform kick in—when people's records move digitally without their being physically present, cash is eliminated from the system, and an individual's consent is secured—that the full potential of India Stack and the power of a platform is unleashed.

In trying to synthesize all of this, a thought occurred to Purohit. Is it possible that while we may claim to have shifted from 'pipes' to 'platforms', we haven't really made that shift? When thought about, a pipe is a supply chain. But then, a platform is a supply chain as well. It supplies you with what you want, but in more sophisticated ways. To get the import of that, some examples are called for.

Consider Amazon, for instance. If you choose to buy something, its algorithms make a quiet note of it. Over time, much like salesmen in a store you frequent, these algorithms understand your preferences, and the 'Recommendation Engine' suggests something it 'thinks' you may like. Of the $136 billion Amazon earned as revenue in 2016, close to 48% ($47.6 billion) was generated,[19] in part because the Recommendation Engine works well. It induces people to buy more.

For that matter, take Facebook. If it were a country, it would be as large as India and China put together. In 2017, it had more than 2 billion people active on it.[20] Intriguingly, 241 million people on it are from India[21]—and the numbers are still growing. That outstrips the 214 million Americans on it. Into these, extrapolate the fact that in 2016 alone, the company generated close to $27 billion in revenue.

That underscores the point that when data is collected in humungous quantities, it can be used to influence behaviour. That is why commercial platforms are under much scrutiny in many parts of the world. If popularity and revenues are metrics to go by, entities like these are wildly successful ones.

What if a metaphor from biology be embedded, Purohit wondered. Biological systems are complex and dynamic, and constantly adapt to embrace diversity. Contemporary biological history has one such example. Some 540 million years ago, a Cambrian Explosion—an evolutionary burst that led to great diversity in life forms—occurred. This is recent in biological time because there is evidence the first form of life appeared 3,800 million years ago. The after-effects of this were radiations of biological diversity that lasted 70–80 million years. This is an interesting metaphor.

While oxygen was around for almost 2,800 million years, it was a scarce resource because most of it was buried. During the Cambrian explosion, though, oxygen became abundant. This gas catalysed the emergence of new life forms.

When this metaphor is placed in an Indian context, education is scarce. Much the same can be said of sanitation, healthcare, and potable drinking water. The question to ask then is, what is it that can be infused into the system to create an environment where all that seems scarce now becomes abundant?

The team brainstormed and figured a few things. Basis their learnings from implementing Aadhaar, they now knew many people wanted it for one reason. For them, establishing their identity was either impossible or difficult. When it was solved, other problems were solved as well. To prove how, 'Reference Applications' were created.

For instance, the problem is most acute at the grassroots. Reports from Karnataka in April 2017 suggested that 1 million farmers[22] had been hurt by one of the worst droughts in 100 years.[23] The state government requested for funds from the National Disaster Relief Fund. By March 2017, Rs 6.71 billion was credited directly by way of subsidies into the accounts of those who had bank accounts seeded with Aadhaar numbers.

In an earlier time, this would have been a laborious process. While there is much quibbling between the Centre and state over the quantum of funds, there is no taking away from the fact that the people got the funds. This was a 'Reference Application' that propelled an abstract idea like identity and Aadhaar into the public imagination.

The larger point here is that it proved to people who benefited that there was a tangible reason for them to get an identity and become a part of the formal sector. From a citizen's perspective, it provided them with a direct benefit; the government got a better grip on its expenses; and the private sector felt compelled to innovate so that they may serve those who had entered the formal economy. The proverbial 'fortune at the bottom of the pyramid' was now beginning to look tangible.

'Certain sections of India have gone to a higher equilibrium and cannot go back to a lower equilibrium,' offers Maruwada by way of perspective.

This is the learning that emerged, says Maruwada: 'We cannot create one solution that we manufacture for someone to use. Because, especially in social problems, it involves their incentives and capabilities. All this exists in economies like India that is performing at a low equilibrium ... We have to migrate to the higher equilibrium. That is our theory of change. And Societal Platform is our approach to meet this challenge.'

Inducing Change

With Project Aadhaar, the team started out with a stated mission—provide identity. Who was to know what may emerge from it? Soon, it was used for direct benefit transfer in place of cooking gas subsidies. And it reinforced an idea—the Diffusion of Innovation Theory.

As a thumb rule, innovators and early adopters comprise roughly 16% of the population. Roughly, about 68% of the population wait and watch. Of these, half will come to the party early, while the other half will take a little while. But come they will. The 16% that remain will be laggards or naysayers. Expending energy on them is pointless.

To get the attention of the innovators and the early adopters, it is important to pique their curiosity. When looked at from this lens, and basis the learning from implementing Aadhaar and India Stack, the question was, what Reference Application could they think up to induce change in education?

It had to be done thoughtfully. 'You can induce labour. You cannot engineer labour. You can engineer a car. You cannot induce a car. It is a complete departure from the mechanistic world view that dominates current platform thinking,' Maruwada explained.

As theories go, it sounds stimulating and good. But there are two issues with complex adaptive systems and Societal Platforms that emerge out of these thoughts. The outcome that may emerge is something you have no visibility around or control over. Because if a platform is complex and dynamic, it is not possible to imagine the future, for multiple futures exist. How do you convey the idea?

Here again, Maruwada has an interesting take. There are risks involved in not doing anything and not challenging the status quo, especially if your eye is trained to look only at the negative outcome. A good example of which is Aadhaar. Does it compromise privacy? 'It is a risk. But not doing Aadhaar is the equivalent of a ship staying in the harbour when its passengers are desperately trying to go somewhere else.'

Maruwada then goes on to embed some questions to which there are no clear answers. All he does is suggest that it will have to emerge.

'Are all unintended consequences the outcomes of the actions of one actor? If it is, who is that actor? The natural instinct of mankind is towards progress. When railways were created, that cattle got killed was unintended. When highways were introduced, people got killed. When electricity was invented, people got killed. We weren't around then. But people may have had their doubts. They were legitimate ones. But ought we have stopped railways in their tracks, highways from getting built, or electricity from being transmitted? I am unclear in my mind at what point does someone say, it is worth the risk, let us go ahead. But if a group of people come together and say this is how we can solve all problems, and the group keeps expanding exponentially, then it is not the power of an individual, but a system at work and society co-creates.'

That his argument is not on paper alone is evident now because the work that was being put in on the platform led by Purohit got the attention of the government of India. Is it possible, they asked, to create something that can be rolled out across India, but with enough flexibility built into it that allows states to customize it for their own requirements?

The team at EkStep said it is possible—and it took 6 months from the time it was thought up for the NTP to go live. 'Because it was the government of India, there was scale. We did not tell them to do it. But because the government of India needed it and wanted to do it, there is a higher chance of sustainability,' says Maruwada.

'That is the power of digital infrastructure. It enables deployment of education in a specific context,' says Maruwada.

As for DIKSHA, the government has released a strategy document on how to go about it. But it is not a guidebook. Instead, it is a manual that contains the five design principles that work best on this platform.

As for the traction, now that the Finance minister has made it clear in his speech, Maruwada and Purohit say the bureaucracy will push their pedals on it. It is an idea that has now gone beyond the founding team. That, in any case, they say, was the intent.

When pressed on how to measure the outcomes, Purohit suggests some questions be asked. 'When you talk about outcomes, you must understand that there is not just one person, but many people who are responsible for it. To give you an example, if you look at the GPS how are you going to measure the return on investment on it? Is Google an outcome of GPS? Or are the revenues of Google an outcome of GPS?'

He goes on to point out that if we look around the landscape that is the Internet and attempt to capture the GDP it generated, how may we even begin to do it? 'If Uber were to build a GPS on the other hand, it would build one that generates revenues only for Uber, not for the entire ecosystem. That is what I mean, that in trying to build Societal Platforms, you have to think very differently. You must learn to collaborate.'

It is when they were pushed into this mode of thinking, says Maruwada, that they came up with QR codes for textbooks that are now part of DIKSHA. 'Our mission was literacy. The technology can be embedded into a textbook. But tomorrow, it can be put on a bag of urea for a farmer to use,' he explains.

The key question, he says, is that in trying to find solutions to a problem, where do the answers lie? In technology, in policymaking, or at some intersection? If it is technology, then who may the right partner be? If it is policy, then it requires a different skill set altogether. Deploying the solution at the last mile comes with its own set of challenges and its resolution can come from those working at the grassroots, like NGOs.

To implement DIKSHA on the NTP, for instance, the QR code only offers the technology to discover content. To create compelling content requires a creative content creator and curator. This means letting go of what has been originally thought up and letting a narrative emerge.

But it is a long journey, much like Aadhaar was. It cannot be prescribed, and the platform cannot be engineered, because the principle on which it is built is that it ought to be relevant in a given context. The solution, to that extent, is context specific. It has to evolve.

What all this does is that it opens the doors for India to offer the world a new narrative. The American story was built with America, for America. The Chinese narrative is a hegemonic one that leaves many diplomats unsettled. As opposed to that, India comes across as benign.

Then there is Project Aadhaar that has been implemented at scale and speed across ecosystems because India is a diverse country. The philosophical underpinnings of Aadhaar are very different from the American and Chinese ones. It does not explicitly seek to profit nor does it seek to impose hegemony.

To the contrary, it was built to weather constraints that many developed and developing economies face. For instance, how do you keep costs dramatically low where either fiscal deficit is par for course, like in a Western economy, or it is hard to come by, like in some parts of South Asia or Africa?

The Indian government does not have the institutional bandwidth anymore within the Indian Foreign Service (IFS) cadre to build soft power. As things are, the American diplomatic corp has about 30,000 people manning it and spread across the world. The Chinese are upping the ante and now have over 5,000 of them. The IFS has only 900 people in it—about the same as Singapore does.

What nobody is willing to come on the record to discuss or even talk about unofficially is that there seems to be a tacit understanding among political parties across the spectrum about Nilekani's argument that the government align with iSPIRT, another not-for-profit he is involved with, to spread the gospel of India Stack. There are pointers in history on why there may be a tacit understanding and policy wonks are all too aware of it.

It can be traced to the time when an idea was originally thought up by DARPA, an agency of the United States Department of Defence. This agency is mandated to identify and develop emerging technologies for its military forces. The idea was to create an interconnected network (or the Internet as we call it now), and was first conceived in the 1960s, and called Arpanet. The stated intent was to connect computers across various institutes to share resources and make best use of idle computing time. Over time, protocols on how to use it emerged.

For instance, naming a site is based on the protocol that was set in place by an entity called the ICANN. It is now a not-for-profit consortium and has representatives from 111 countries. What often goes unsaid and unreported is that this too was formed by the American government. It was only on 1 October 2016 that the US Department of Commerce formally ended its association with the entity.

When the Arpanet was built, and as it morphed into the Internet, who could have imagined then what would come of it and how it may impact the world? Over time though, because standards on how to use it had been created, and it was opened up to technologists and hobbyists of various kinds, a 'killer app' emerged—email. That made it a compelling proposition for the masses.

Circle back now to the narratives that are Facebook, Amazon, Netflix, Google, and Uber. As commercial entities, these companies are clear in their pursuit of profits. But they are built on infrastructure created by the American government. There is much evidence to prove that the American government and these entities are aligned as well. Take the battle for net neutrality.

There is a school of thought that argues that access to the Internet must be equitable and democratic for all entities. But this debate has been an ongoing one—and as recently as December 2017, FCC, America's top media regulator voted against it.[24]

This is the sixth time the American government has voted to protect the interests of large American entities. This is because big business maintains that access to their services across the infrastructure that is the Internet must be faster for their users because they can pay for the upkeep of the infrastructure. If implemented, it can have serious implications on businesses not just in the US but also in other parts of the world, and will place the start-up ecosystems at a disadvantage. This, because it will be unviable for them to compete with businesses that can pay the higher prices a regulator may choose to set for faster access to their homepages on the underlying infrastructure. Entities like Facebook, Amazon, Netflix, Google, and Uber can afford it though.

It is a narrative that makes the authorities in China wary. But China is fundamentally different from America in a few ways. To start with, it is an authoritarian state. Unlike in a democracy where parties are elected to power, orders are imposed top-down. Then there is the fact that it is home to over a billion people. That makes it an incredibly large market no government or enterprise whose stated intent is to earn profits can ignore.

And finally, the country has made its intent clear. China wants a Chinese narrative to dominate the world. It will not toe the American line. When looked at from Beijing, it resides at the centre of the world. More importantly, the State owns the Bazaar, and Society must follow orders. That explains why

China had the audacity to shut its doors to American Internet businesses and worked at crafting a new narrative.

This work started a long time ago though. Between 2003 and 2013, China extended $75 billion in loans to various African countries. In Latin America, Chinese diplomats made significant inroads—so much so that even the Nicaraguan Congress allowed a Chinese company to build a canal through the country.[25] And one of the biggest stories unfolding in Asia now is the One Belt, One Road Initiative[26] that is being funded by the Chinese government. This means funding the construction of a physical road on the one hand and recreating the ancient Silk Route on the other. Unveiled in 2013, the plan has the audacity built into it to encompass 60 countries and is budgeted to cost anywhere between $4 trillion and $6 trillion. The timeline is not known, because multilateral partnerships have to be put in place.

Edward Tse puts all of this into perspective in his 2015 book *China's Disruptors*.[27] To do that, he coined a term—SOOT, an acronym for Scale, Openness, Official Support and Technology.[28] 'These factors', he writes, 'can be best seen operating together in China's growing digital economy marked by huge people participation, transparency, supportive government policies and, well, technology.'

On the back of initiatives like these, while China evokes shock and awe on the one hand, its soft power has been on the rise as well. It made its debut and ranked 30 in 2015 on The Soft Power 30, a scale developed by Portland, a strategic communications consultancy that works with governments, businesses, foundations, and NGOs across the world. In 2016, it moved up the scale to 28. By the end of 2017, it had moved to 25.[29]

How did this happen? An overview posted by Portland offers clear pointers to where it is headed:[30]

> The opening of hundreds of 'Confucius Institutes', combined with extensive international branding initiatives, have only strengthened China's cultural offering. And as China continues to post strong performances in innovation and R&D spending, its efforts have led to the increasing global influence of Chinese brands like Huawei and Alibaba.
>
> Politically, China is in a strong position to shoulder greater global responsibility as America turns inwards and distances itself from free trade and climate commitments …
>
> However, gains made in its soft power standing are somewhat undermined by China's hard-line approach to foreign policy and human rights … increased defence spending, and ongoing construction in the South China Sea have led to poor polling performances in other Asian countries.
>
> However, China's overall improvement in the polling data, and its notably high scores in Africa, indicate there is opportunity for China to adopt a more favourable global position.

India does not figure on the list.

But if the history of ICANN and its influence is anything to go by, it is also possible to imagine a world where something that emerged from EkStep resonates with the Indian government. If the world needs an alternative narrative, this can be a good place to initiate a conversation to imagine a coalition that works towards building the future of education at scale, with protocols and standards in place.

To that extent, this may just be the beginning of an India narrative if India must assert its soft power. China hasn't said much about India Stack. It has insinuated though that there is some concern in the government about three countries in particular that are seen as close to China—and is keeping a close eye on how people there are viewing it and what kind of questions are emerging from there.

While volunteers who work for India Stack claim it is an open platform and open to anybody in the world, some alarm bells have been sounded as well to suggest access to India Stack be limited for these countries. Rather, that much thought be given to where it is deployed, and that it must be done in consultation with the highest offices of the country, including the PMO.

But there is a thorny right wing narrative running through India that the PMO must balance delicately on the one hand to safeguard expressions of interest in the project from the Middle East that have large Muslim populations. Then there are right wing narratives in other parts of the world, Europe included, that do not like the sound of a foreign narrative entering their national discourse. The way things are, a story from India may not go down well in either of these regions.

But as Maruwada and Purohit suggest, the world is a complex, dynamic system that no one has wrapped their heads around. All these are theories. And theories are abstracted from experiences in the past. To that extent, their submission is that Societal Platforms be seen as a codification of guard rails they are putting out in the public domain for the world to consume and contribute to.

Who knows what outcomes may emerge out of it? The journey will outlive the founding team and the outcomes may be of the kind we have not imagined. Who would have imagined that a project to deliver a unique identity for those who reside in India would have evoked a cacophony of protests, catalysed solutions in domains they hadn't imagined, or for that matter set the rules for a new geo-political game?

Notes

1. PIB, 'Union Budget 2018: Full Text of Arun Jaitley's Budget Speech', *Mint*, 1 February 2018, https://www.livemint.com/Politics/6ZTmv653VqU5ghPAcWCfTJ/Union-Budget-2018-Full-text-of-Arun-Jaitley-budget-speech.html, viewed on 10 August 2018.
2. NTP, 'National Teacher Platform, Strategy and Approach', Ministry of Human Resources and Development, May 2017, http://mhrd.gov.in/ntp/doc/NTP_Strategy&ApproachPaper.pdf, viewed on 9 July 2018.
3. FIS, 'Flavors of Fast 2017: A Trip around the World of Global Payments', 2017, https://www.fisglobal.com/flavors-of-fast-2017, viewed on 9 July 2018.
4. BBC News, 'South Korean ID System to Be Rebuilt from Scratch', 14 October 2014, http://www.bbc.com/news/technology-29617196, viewed on 9 July 2018.
5. *Financial Times*, 'Japan to Adopt Wide Ranging National Identity System', 3 September 2015, https://www.ft.com/content/8526da22-5205-11e5-8642-453585f2cfcd, viewed on 17 July 2018.
6. *The Mainichi*, 'Everything at Your Fingertips: Govt to Test Biometric ID-Based Shopping System', 15 September 2016, http://mainichi.jp/english/articles/20160915/p2a/00m/0na/012000c, viewed on 9 July 2018.
7. *Finextra*, 'EU Opening Is Not Just for Banks', 22 May 2018, https://www.finextra.com/blogposting/15129/eu-open-banking-is-not-just-for-banks, viewed on 9 July 2018.
8. Angelica Mari, 'PSD2 and the API Challenge for Open Banking', *Diginomica*, 2 March 2018, https://diginomica.com/2018/03/02/psd2-and-the-api-challenge-for-open-banking/, viewed on 9 July 2018.
9. Medha Basu, 'Singapore's New Digital Identity Scheme', *GovInsider*, 3 March 2017, https://govinsider.asia/digital-gov/

singapore-trials-new-digital-identity-scheme/, viewed on 9 July 2018.

10. Jeanette Rodrigues, 'India ID Program Wins World Bank Praise Despite "Big Brother" Fears', *Bloomberg*, 16 March 2017, https://www.bloomberg.com/news/articles/2017-03-15/india-id-program-wins-world-bank-praise-amid-big-brother-fears, viewed on 9 July 2018.
11. George Akerloff and Rachel E. Kranton, *Identity Economics: How Our Identities Shape Our Work, Wages, and Well-Being* (Princeton: Princeton University Press, 2011).
12. Robert Shiller, 'Stuck in Neutral? Reset the Mood', *New York Times*, 31 January 2010, https://www.nytimes.com/2010/01/31/business/economy/31view.html, viewed on 9 July 2018.
13. Amrit Raj and Upasana Jain, 'Aadhaar Goes Global, Finds Takers in Russia and Africa', *Mint*, 9 July 2016, https://www.livemint.com/Politics/UEQ908E08RiaAaNNMyLbEK/Aadhaar-goes-global-finds-takers-in-Russia-and-Africa.html, viewed on 9 July 2018.
14. Meera Srinivasan, 'Sri Lanka Is Keen to Introduce an Aadhaar-Like Initiative', *The Hindu*, 21 December 2017, http://www.thehindu.com/news/international/sri-lanka-is-keen-to-introduce-an-aadhaar-like-initiative/article22099494.ece, viewed on 9 July 2018.
15. Kai Schultz, 'Sri Lanka, Struggling with Debt, Hands a Major Port to China', *New York Times*, 12 December 2017, https://www.nytimes.com/2017/12/12/world/asia/sri-lanka-china-port.html, viewed on 9 July 2018.
16. Abhishek Waghmare, '6 Indicators of India's Looming Demographic Disaster', *IndiaSpend*, 2 May 2016, http://www.indiaspend.com/cover-story/6-indicators-of-indias-looming-demographic-disaster-99797, viewed on 9 July 2018.
17. UNDP, Asia-Pacific Human Development Report, 2016, http://www.asia-pacific.undp.org/content/dam/rbap/docs/

RHDR2016/RHDR2016-full-report-final-version1.pdf, viewed on 9 July 2018.

18. Dominic Barton, 'Capitalism for the Long Term', *Harvard Business Review*, March 2011, https://hbr.org/2011/03/capitalism-for-the-long-term, viewed on 9 July 2018.
19. Shabana Arora, 'Recommendation Engines: How Amazon and Netflix Are Winning the Personalization Battle', *MTA Martech Advisor*, 28 June 2016, https://www.martechadvisor.com/articles/customer-experience-2/recommendation-engines-how-amazon-and-netflix-are-winning-the-personalization-battle/, viewed on 9 July 2018.
20. *Statista*, 'Number of Facebook Users Worldwide as of 1st Quarter 2018', https://www.statista.com/statistics/264810/number-of-monthly-active-facebook-users-worldwide/, viewed on 9 July 2018.
21. *Statista*, 'Leading Countries Based on Number of Facebook Users as of April 2018', https://www.statista.com/statistics/264810/number-of-monthly-active-facebook-users-worldwide/, viewed on 9 July 2018.
22. K.R. Balasubramanyam, 'In a First, Karnataka Uses Aadhaar to Transfer Subsidy to One Million Farmers', *Economic Times*, 14 March 2017, https://economictimes.indiatimes.com/news/economy/policy/in-a-first-karnataka-uses-aadhaar-to-transfer-subsidy-to-one-million-farmers/articleshow/57635641.cms, viewed on 9 July 2018.
23. Janaki Murali, 'Karnataka Faces Worst Drought in 42 Years', Firstpost.com, 4 April 2017, https://www.firstpost.com/india/karnataka-faces-worst-drought-in-42-years-can-siddaramaiah-save-bengaluru-before-taps-go-dry-3367460.html, viewed on 9 July 2018.
24. Dominic Rushe and Lauren Gambino, 'US Regulator Scraps Net Neutrality Rules that Protect Open Internet', *The Guardian*, 15 December 2017, https://www.theguardian.com/technology/2017/

dec/14/net-neutrality-fcc-rules-open-internet, viewed on 9 July 2018.

25. John Jewell, 'China, the USA and "Soft Power"', Cardiff University Journalism, Media and Culture Blog, 9 November 2013, http://www.jomec.co.uk/blog/is-american-culture-thriving-in-china/, viewed on 9 July 2018.
26. McKinsey Report and Podcast, 'China's One Belt, One Road: Will It Reshape Global Trade?', July 2016, https://www.mckinsey.com/featured-insights/china/chinas-one-belt-one-road-will-it-reshape-global-trade, viewed on 9 July 2018.
27. 'Innovation in China: An Interview with Edward Tse', 18 June 2018, https://www.xuehua.us/2018/06/18/innovation-in-china-an-interview-with-edward-tse/, viewed on 26 July 2018.
28. *Economic Times*, 'Why China Produces More Entrepreneurs than India', *Economic Times*, 27 November 2017, https://economictimes.indiatimes.com/news/economy/indicators/why-china-produces-more-entrepreneurs-than-india/articleshow/61817018.cms, viewed on 9 July 2018.
29. US Center on Public Diplomacy, 'The SoftPower 30: A Global Ranking of SoftPower 2017', https://softpower30.com/wp-content/uploads/2018/07/The_Soft_Power_30_Report_2017-1.pdf, viewed on 26 July 2018.
30. US Center on Public Diplomacy, 'The Soft Power 30: A Global Ranking of Soft Power 2018', https://softpower30.com/wp-content/uploads/2018/07/The-Soft-Power-30-Report-2018.pdf, viewed on 26 July 2018.

Epilogue

MOST OF US HAVE GOT USED TO BEING OFFERED solutions on a platter. That is why we find it difficult to wrap our heads around the idea of platforms, or (digital) Lego blocks, if you will. By thinking of identity as a platform—much like the GPS, which acted as a platform that led to more innovations—Nandan Nilekani and the founding team at UIDAI nudged us into a new world. How the government, businesses, and social sector use this crate of Lego blocks is really up to them.

Our journey into this new world started with Aadhaar. Very soon, other pieces of digital infrastructure—such as eKYC, eSign, DigiLocker, UPI, Consent Layer—emerged as part of India Stack, allowing for 'presenceless', paperless, and cashless solutions.

And that's only the story so far. The team that worked on India Stack is at work on a Health Stack, which promises to kick-start innovations in the healthcare industry, and a Travel Stack, which aims to make travel more convenient. It is only a matter of time before more such stacks are developed by those who are inspired by Aadhaar and India Stack.

How will it pan out?

The story of Aadhaar offers us pointers to a few principles. These are important for strategists, business leaders, policymakers, technologists, civil society and everyone who aspires to build a better tomorrow.

#1 The world is getting messier by the nanosecond, and leaders need to figure out how to operate in it.

The world is complex, dynamic, and driven by human beings whose motivations, decisions, and actions we sometimes don't understand. That's probably why social science, which sets out to demystify the world, fails so often: much of the world resists decoding.

Add to this the non-linear complexity of innovations—the rise and rise of exponential technologies such as data analytics, artificial intelligence, robotics, and the Internet of Things, among others.

In this world, data is the new black gold. Every device—not just phones and computers—is designed to extract granular data about the user, with an implicit promise of making their lives better by giving them a hyper-personalized experience. It is inevitable that governments and businesses drool over the multiple possibilities this promises. But there are also risks that need to be addressed.

The intense arguments that followed the release of Justice Srikrishna Committee report and the draft Personal Data Protection Bill in July 2018 point to the difficulties of arriving at a consensus. Strategists today must not just operate but also thrive in this complex and messy world to earn their pay cheques.

#2 Great ambition attracts great opposition. Leaders have to pick battles wisely.

Across the world, business leaders are less ambitious than before. They are more suspicious of long-term vision. Case in point: the popularity of terms like 'fail fast, fail often', 'agile', 'pivot', etc.

To solve some of the biggest problems—poverty, illiteracy, unemployment, diseases, climate change—the world needs ambitious leaders. But big ambitions also attract intense opposition, often from unexpected quarters. This takes its toll—emotionally and mentally. And no amount of B-school training or life-coaching can truly steer you through.

It takes wisdom to know when to compete, when to co-operate, and when to take opponents head on—and often, these learnings come at a cost.

Nilekani and his team often picked the right battles: the fight with the Home ministry had to be fought and won. They sometimes picked the wrong battles: they could have avoided their clashes with development activists and instead looked at ways to cooperate with them. Sometimes, they also reacted in the wrong way: their engagement with technology activists left much to be desired.

#3 Policymaking is difficult; exponential growth in technology has made it more so.

One of the ironies of the draft Data Protection Bill, and even the Aadhaar law, which was passed in 2016, is that it was put in place *after* Aadhaar was implemented. Well-thought-out regulations can minimize disruptions, and that's why laws to govern programmes are crafted *before* those programmes are rolled out. Yet, disruption was inevitable in the case of Aadhaar. Much of it has to do with the fact that technology usually outpaces the fastest governments.

This is a phenomenon that Samuel Arbesman, author of two books and scientist-in-residence at Lux Capital, the venture capital firm that invests in technologies, has studied extensively. Arbesman's research shows that the knowledge a doctor gains in medical school becomes obsolete in 45 years. This number reduces drastically for those who hold public office and are engaged in policymaking: unless they work differently, the amount of data being churned out can render them obsolete in under five years.[1]

To be future ready, India needs to:

- frame public policies with teams that include technologists, data scientists, micro-economists, lawyers, and ethicists.
- build cybersecurity infrastructure, create a pool of cybersecurity warriors, and introduce frameworks for cybersecurity insurance, among other things.[2] Singapore has a minister for cybersecurity, signalling its seriousness about safeguarding its data. Even now, UIDAI lacks a chief security officer.
- develop a culture of designing better governance solutions and promoting data literacy, as mere data regulation is no longer sufficient.

#4 Technology is essential to solve problems, but a system's capacity to absorb it and its intent to solve a problem matters much more.

Many technologists across the world believe that all problems can be solved by technology.

By making it techno-reliant, Aadhaar shifted the government's approach to solving problems. Instead of building an end-to-end solution, it built a platform. This was much better than thrusting cookie-cutter solutions built in Bengaluru on Budaun or Bagdogra. However, it wasn't free of problems either. Ultimately, we can't solve problems by technology alone. The *capacity* of the system and the players' *intent* to solve the problem matter equally, if not more. Technology can enhance that capacity. Often, it can do nothing about the intent.

The state governments and district administrations made several mistakes, many of them basic, such as not having a backup for biometric authentication. Inevitably, it was the citizens who took the hit.

But then, the world is a messy place and there is no such thing as the perfect time. Plus, it is rather naive to underestimate the ability of local administrators and technologists to create solutions best suited for their constituencies. In trying to micro-manage everything, one ends up killing people's ability to take initiative. The trick lies in finding the right balance.

#5 Activism is necessary; but 'echo chamber activism' can undermine the larger cause.

The right medicine, at the right dose, can cure diseases. But if consumed in high amounts, it can harm or even kill. Activists fighting against exclusion, privacy, security risks, and poor implementation are vital for the ecosystem. However, going by the outrage in the media—social or otherwise—it appears many have lost the plot.

Optimists expect these shouting matches to fade away once the Supreme Court judgment comes in, leaving space for constructive dialogue. But who knows if they may ever see one another's perspectives?

The obsession to make Aadhaar perfect has made activists amplify even the small mistakes that cause no harm and can be easily fixed.

This has had three negative consequences:

- The big issues around Aadhaar never get addressed.
- At the same time, the constant noise about Aadhaar and associated entities may lead to a boy-who-cried-wolf-like situation. People may end up dismissing even real warnings as false alarms.
- The different factions seem to engage in dialogue only on television channels and social media, or in courtrooms, that is, spaces where the differences end up getting amplified.

#6 Society must embrace technology and educate itself about the risks involved.

Aadhaar established that Indians understand the power of technology in a deeply intuitive way. In Suravaram, a small village in Agiripalli Mandal, Andhra Pradesh, we spoke to an old lady too weak to get up from her bed, but who drew her pension from a banking correspondent using a micro-ATM (unlike some of her friends, her fingerprints seemed to work). She said, 'Trains got cities closer to villages, and this machine got banks closer to us.' She didn't need anyone to explain what financial infrastructure meant. However, numerous instances of anecdotal evidence suggest that people are not aware of digital risks. They give away their OTPs and share their ATM pins. Sometimes, they gave consent without realizing the risks. Many Airtel customers were shocked to find their subsidies getting delivered to an Airtel Payments bank account that they hadn't applied for and didn't know existed.[3] Also, some of the new features of Aadhaar, such as biometric lock and Virtual Aadhaar, demand a steep learning curve. It's not clear if the market will fix digital illiteracy by itself; a nudge from the government and social sector will surely be beneficial.

#7 Systems evolve through an act–learn–act cycle.

Many of today's problems need urgent resolution—they cannot wait until tomorrow. There's an overwhelming need for action. But since the world is messy, complex, and dynamic, solutions can be effective only when they evolve. The cycle of act–learn–act helps in that evolution. This is evident in the growth of Aadhaar into India Stack, and then into Societal Platforms.

The Future

Peter Thiel, venture capitalist and philanthropist, writes in his book *Zero to One*[4] that he likes to pose this question to founders and potential hires: What important truth do very few people agree with you on? This is the business version of the question: What great business is nobody else building?

Building on this, we ask, what moral issue do we accept as normal, but would shock our grandchildren?

Our answer: inequality.

An economist would argue that the way to get around this disparity is to redistribute wealth, perhaps through Universal Basic Income. A philosopher, however, might contend that humans are creative beings who need to find fulfilment by enhancing their skills to better their lives. If so, the focus ought to be on providing everyone with education, healthcare, and access to finance, along with the opportunities that come with it.

As storytellers, we were excited about Aadhaar because it took a radically different approach to solve this problem. At the risk of sounding repetitive, the founding team did not offer an overarching technology solution to address every need. Instead, they designed Lego bricks that, though useless by themselves, create something different in the hands of each architect. Basically, they left it to the ingenuity of the user.

Now, the world is complex and dynamic. All interventions, policies, and technologies will have multiple consequences—short-term and long-term, intentional and unintentional, good and bad.

The real impact of Aadhaar—if we survive the polarized debates that dominate the current narratives—will depend on whether we use innovation to provide opportunities for everyone tomorrow. And that would depend on what we—governments, large businesses, start-ups, social enterprises, non-profits, freelancing professionals, techies, activists, and society at large—do today.

In the preceding pages, we attempted to document how Aadhaar triggered interesting debates and spurred some far-reaching initiatives that go beyond Aadhaar. The outcomes of these debates are intended to address some of the fundamental issues that India ignored for too long even as the world was speeding ahead.

Aadhaar was in many ways responsible for a case that led the Supreme Court to assert that privacy is a fundamental right. It also led to the setting up of the Justice Srikrishna Committee, which has come up with a draft law on data protection.

We don't know what changes it will undergo before the parliament passes it as law. The point is that we as a country are discussing some of the most important issues around data. What we do with the law could well set the template for the entire developing world.

Aadhaar and India Stack have encouraged businesses to think about Indians in the middle and bottom of the pyramid. In fact, Societal Platforms, an evolution of thinking behind Aadhaar and India Stack, has already showed how the social sector can harness technology for social transformation.

The use of Aadhaar by different government agencies has reiterated that the government is not a monolith. While there are pockets of excellence, there are also examples of bureaucrats who don't understand technology and can thus harm the citizens they mean to serve. It has also pointed to serious issues around cybersecurity. Aadhaar has kindled broader debates on state surveillance, and India might finally be forced to put in place institutional mechanisms and public oversight to keep state surveillance in check.

Aadhaar's rollout and impact made development agencies such as the United Nations and several developing countries take platform-based thinking seriously. Within the country, the cantankerous debates made several activists, academics, and observers introspect about the meaning of constructive participation and the nature of public discourse. The development of the various elements of India Stack under the umbrella of iSPIRT has shown a new way for civil society to engage with public policy and to create digital public infrastructure.

Above all, the innovations around Aadhaar have opened up doors for millions of Indians to embrace the benefits of digital economy, so far enjoyed only by those at the top of the pyramid.

Equally, while some of the actions mentioned earlier were initiated to address the risks posed by Aadhaar, they also address the bigger, broader, and more dangerous risks posed by the digital economy. The big question is whether we are up to the task.

In this age, there are several reasons why we should be cynically pessimistic about the future. Especially one driven by technology. The stories we tell ourselves—through our movies and books—usually highlight what can go wrong. We just have to look around to see the damage it has caused—global warming, depression induced by addictive devices, state surveillance, and so on.

But there are also reasons to be optimistic. In *Factfulness: Ten Reasons We're Wrong About the World*,[5] Hans Rosling pointed out that contrary to popular belief, the world has actually become a better place to live in.

This did not happen by itself; it was propelled by human ambition and endeavour. By innovations. By building the right kind of institutions. And above all, by taking responsibility.

When we started researching this project, one of the first people we met was Manish Sabharwal, chairman and co-founder of Teamlease Services. He was intrigued by the title of the book. 'The Aadhaar *Effect*?! But the impact of Aadhaar has hardly begun. Do you mean the *potential* of Aadhaar?' he asked.

What we do know is that when presented with tough options, we have to make a choice. Having chosen a path, we have to work on it and implement it, striving to impact the most number of people in the best possible way. That is The Aadhaar Effect.

It is a work in progress. We believe this is a project that will be in perpetual beta and we will be better off for it. That is why the world's largest technology project matters.

Notes

1. Charles Assisi, 'The Half-Life of Truth Is 45 Years', *Founding Fuel*, 13 April 2018, http://www.foundingfuel.com/article/the-halflife-of-truth-is-45-years/, viewed on 10 August 2018.
2. National Venture Capital Association, 'NVCA Chair Testifies before Senate Commerce Committee on Venture-Backed Innovation in Cybersecurity', 22 March 2017, https://nvca.org/pressreleases/nvca-chair-testifies-senate-commerce-committee-venture-backed-innovation-cybersecurity/, viewed on 10 August 2018.
3. Surabhi Agarwal, 'Bharti Airtel Likely to Face Penalty for Misusing Aadhaar Details', *Economic Times*, 18 December 2017, https://economictimes.indiatimes.com/news/economy/policy/bharti-airtel-likely-to-face-penalty-for-misusing-aadhaar-details/articleshow/62110386.cms, viewed on 10 August 2018.
4. Peter Thiel and Blake Masters, *Zero to One: Notes on Startups, or How to Build the Future* (New York: Crown Business, 2014).
5. Hans Rosling, Ola Rosling, and Anna Rosling Rönnlund, *Factfulness: Ten Reasons We're Wrong about the World—and Why Things Are Better Than You Think* (London: Hodder and Stoughton, 2018).

Index

C

D

E

F

G

H

I

O

P

About the Authors

N.S. Ramnath is a senior writer at Founding Fuel Publishing, a digitally led media and learning platform. His main interests lie in technology, business, and society, and how they interact and influence one another. A Polestar 2017 award winner, he writes a regular column on disruptive technologies and takes stock of news and perspectives from across the world. He is also involved with the data start-up How India Lives.

Prior to Founding Fuel, Ramnath was with *Forbes India* and *Economic Times* as a business journalist. He has also written for *The Hindu, Quartz*, and *Scroll.*

He tweets from @rmnth and spends his spare time reading on philosophy.

Photo credit: Vikas Khot

Charles Assisi is a co-founder at Founding Fuel. He writes a column in the business daily *Mint.* He was the managing editor at *Forbes India* and the founding editor of *Forbes Life India*. Charles earned his spurs when he introduced *CHIP*, the

Munich-based technology magazine, to India. This assignment led him to work in newsrooms in Europe and other parts of Asia. Subsequently, the *Times of India* invited him to take over as its national technology editor and later as its national business editor.

His work earned Charles the Polestar and Madhu Valluri Awards.

Charles tweets from @c_assisi.

About Founding Fuel

Founding Fuel was formed by Indrajit Gupta, C.S. Swaminathan, and Charles Assisi as a platform for ideas, insights, practices, and wisdom essential to build the enterprise of tomorrow.

Indrajit Gupta was founding editor of the India edition of *Forbes* magazine and resident editor at *The Economic Times*. He is also a columnist, entrepreneurial mentor, and speaker at industry conferences. C.S. Swaminathan is focused on customer engagement, marketing, analytics, and technology. In earlier avatars, he led the consulting and technology practices at global entities in India and North America.

The Aadhaar Effect is the first in a series around themes that Founding Fuel has identified as crucial for those in leadership roles. To accomplish this, it has engaged in formal relationships with fine minds and universities across the world. The co-founders believe this is one way that learnings from thought leaders can be distilled and transmitted on what it takes to shape public policy and build enterprises that last.